DEDUCING FUNCTION FROM STRUCTURE

VOLUME 2

INFORMATION PROCESSING IN THE RETINA

DEDUCING FUNCTION FROM STRUCTURE

FRITIOF S. SJÖSTRAND
Department of Biology
University of California at Los Angeles
Los Angeles, California

VOLUME 2

INFORMATION PROCESSING IN THE RETINA

ACADEMIC PRESS, INC.
Harcourt Brace Jovanovich, Publishers
San Diego New York Berkeley Boston
London Sydney Tokyo Toronto

This book is printed on acid-free paper. 

Academic Press, Inc.
San Diego, California 92101

United Kingdom Edition published by
Academic Press Limited
24–28 Oval Road, London NW1 7DX

Library of Congress Cataloging-in-Publication Data

Sjöstrand, Fritiof S. (Fritiof Stig), Date.
Deducing function from structure / Fritiof S. Sjöstrand.
p. cm.
Includes bibliographies and indexes.
Contents: v. 1. A different view of membranes -- v. 2. Information processing in the retina.
ISBN 0-12-647655-1 (v. 1 ; alk. paper). -- ISBN 0-12-647656-X (v. 2 : alk. paper)
1. Structure-activity relationships (Biochemistry) 2. Membranes (Biology) 3. Retina. I. Title.
QP517.S85S56 1989
574.87'5--dc19 89-99
CIP

Printed in the United States of America
90 91 92 93 9 8 7 6 5 4 3 2 1

To Ragnar Granit whose work directed my interest to study the retina.

Preface

Electron microscopy has created entirely new possibilities for studying the nervous system by revealing the patterns according to which neurons are connected, thereby establishing the basis for a new approach—neurocircuitry analysis—and extending this analysis far beyond classical neuroanatomy at both the light microscopic and the electron microscopic levels. Examples of primary events in information processing, events that are not accessible for recording by means of the electrophysiological technique, can now be deduced on the basis of the patterns of neural connections and a knowledge of the functional characteristics of the participating neurons.

The various types of responses recorded electrophysiologically from bipolar cells are explained on the basis of such deductions, revealing the basic principles of "on-," "off-," and "on/off-" responses as recorded from ganglion cells.

The complex circuitry discovered in the outer plexiform layer shows that this layer is competent for rather complex processing of information at the photoreceptor level.

A particular type of neural connection offers a simple explanation for spatial brightness contrast enhancement and for the duplicity of vision without having to assume any basic difference in the function of rods and cones.

Structure has played a major role in the deduction of function, and many examples of deductions facilitated by the logic of the design of the neural circuitry are presented.

I extend my thanks to Professor Frederick Crescitelli for the many valuable discussions that I have enjoyed and profited from over the years.

F. S. Sjöstrand

Table of Contents

DEDUCING FUNCTION FROM STRUCTURE

VOLUME 2

INFORMATION PROCESSING IN THE RETINA

Chapter 1: An Approach to the Analysis of the Basis for Information Processing in Neural Centers

1. The Integrative Activity of the Nervous System

The central nervous system receives information from a number of sense organs through afferent nerves. In the central nervous system, this information is processed by multistep integration. As a consequence of this integration, efferent signals are transmitted to various tissues in the body, exerting a control of the function of these tissues.

An enormous amount of information concerning functioning of the nervous system has been collected. However, it is important to establish what we really know about the following basic functions: information transmission from sense organs to the central nervous system, information processing within neural centers, and signal transmission from the central nervous system to effector organs and tissues. We then find that much is known about the first and last aspects of the function of the nervous system but that very little is known about how information is processed. In fact, there is not a single information processing center in the nervous system the integrative activity of which has been revealed to the extent that the basic neural interactions associated with processing of information in the center are known. We only know the behavior of certain neurons that offer conditions favorable for the recording of their activity. The relationship of mathematical models developed by cyberneticists to the real events in the nervous system cannot be established. This gap in our knowledge appears particularly severe when we consider that information processing is the basic function of the nervous system. The reason for this lack of knowledge is that the techniques applied to the analysis of the nervous system have serious limitations and so have contributed only fragmentary information. Let us examine these limitations.

Signal transmission and information processing involve the participation of discrete cells, *neurons*, with the proper shapes for receiving information and for sending signals over distances of varying length. The neuron consists of a cell body from which a more or less elaborate tree of dendritic processes extends, offering a large area for contacts with other neurons, including sensory cells supplying the input to the neuron. One process, the axon, extends from the cell body as the structure by which signals are sent out by the neuron. The axon branches and can thereby contact a large number of other neurons or effector cells.

The neuron illustrates some basic principles of the organization of the nervous system in general. The dendritic tree, in which a number of branches converge toward the cell body, illustrates the convergence that characterizes the input side of a nervous center. The axon, with its splitting up into a number of end branches, illustrates divergence by which the same message can be transmitted to a large number of cells.

Anatomical analysis of the organization of neurons in the nervous system has been based on the technical approach pioneered by Ramon y Cajal (1892, 1894). By impregnating the cells with a heavy metal according to the Golgi technique, this approach reveals to a certain extent the shapes of individual neurons. A neuron appears dark in the light microscope, and the massive precipitation of metal makes even rather thin branches of the neural processes visible under the light microscope. It is, however, never possible to determine to what extent the neuron has been impregnated.

Because neurons are stained very dark, the shape of a neuron can be revealed only when surrounding neurons are unstained. The analysis is therefore confined to the establishment of the shape of *individual* neurons, and contacts between neurons can be revealed to only a limited extent.

In principle, this situation applies also to electron microscopy of neurons stained according to the Golgi technique (Blackstad, 1965; Stell, 1965). Some types of contacts, however, can be identified with respect to type even though the contacted neuron remains unstained. A further weakness of the method is that structural details at the synapses are ruined by the impregnating metal.

Neuroanatomical analysis has revealed some basic features of the organization of the nervous system. It has shown that the structural unit in the nervous system is the neuron. It has allowed one to distinguish between neurons with very different shapes. Other morphological techniques have revealed long-range connections between integrative centers. This structural analysis has furnished the basis for the physiological exploration of the nervous system.

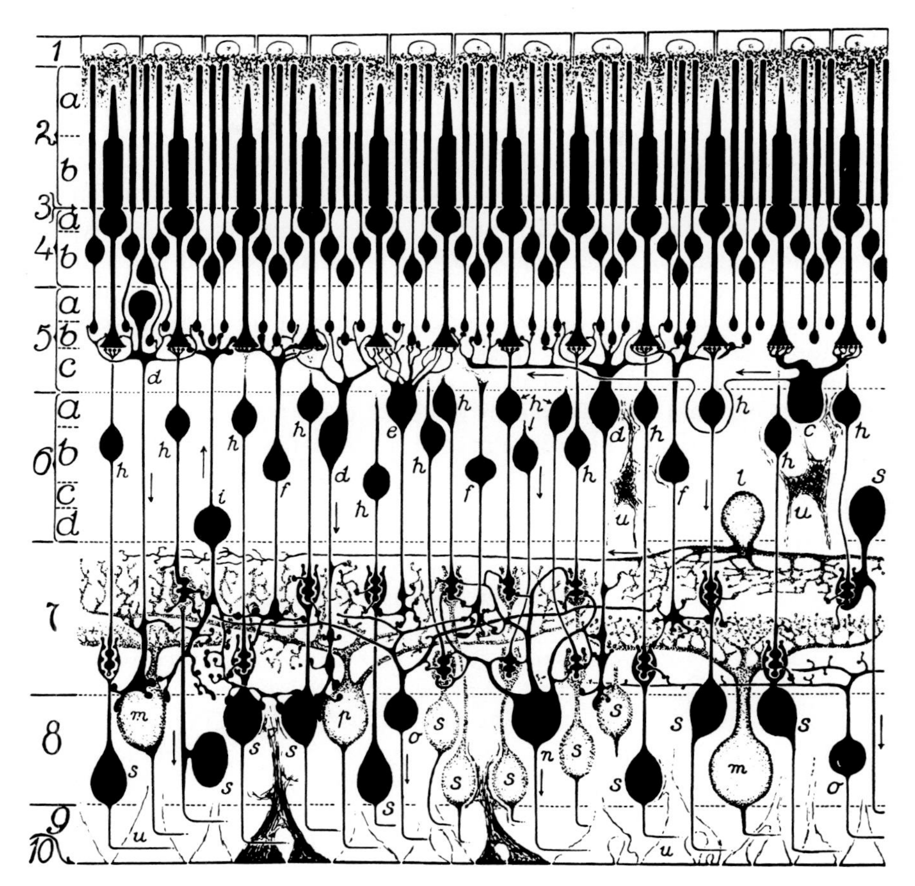

Fig. 1.1. Composite drawing showing the types of neurons identified in the retina by staining individual neurons according to the Golgi technique. This technique reveals the shape of the neurons that can be used for taxonomic classification. Some types of connections between neurons are also illustrated. Such composite drawings do not reveal any real patterns of neural connections. The arrangement of neurons is arbitrary except for correct location with respect to the layers of the retina. It is therefore not possible to deduce the patterns of neural interactions underlying the ganglion cell responses recorded as output from the retina. From Polyak (1941).

On the basis of structural analysis, composite diagrams illustrating the various types of neurons in the retina as well as certain types of connections among these neurons have been published, such as the one in Fig. 1.1 from Polyak's monograph (1941). Such diagrams should not be confused with real circuit diagrams or wiring diagrams. The arrangement of the neurons in the diagram is determined primarily by the author's attempt to include as much information as possible within the smallest possible area. As a consequence, we cannot deduce from the diagram how the neurons interact to shape the output signals from the retina.

Physiological analysis of neurons involves the recording of their electrical responses either extracellularly or intracellularly. This technique is limited in resolution by the size of the microelectrodes that must be inserted into the tissue. For intracellular recording the cell bodies of the neurons must be larger than a certain minimal size, and intracellular recording from individual dendritic branches is not possible. Thus this technique can be applied only to neurons exceeding a minimum size and therefore can contribute only fragmentary information.

Neurons that transmit signals over long distances are usually larger than this minimum size, while in integrating centers the neurons that transmit signals over short distances are smaller, frequently too small to be accessible for intracellular recording. Other neurons within such a center, however, can be large enough to make recordings possible. A rigorous analysis requires, however, that the activity of *all* sizes and types of neurons be recorded. To reveal the timing of the neural interactions, the recordings must be made simultaneously from several neurons.

The limitations of the electrophysiological technique are well illustrated in the case of the retina. A variety of responses have been recorded from various neurons in the retina, and this has led to the collection of descriptive information regarding the way the membrane potential changes under different conditions of stimulation by light. What determines the shape of these responses, however, has not been revealed. One reason for this inability is likely to be that the responses are the consequence of neural interactions at a level that is not accessible for electrophysiological recording.

We find that fragmentary information regarding the structure of neural centers is combined with fragmentary information regarding physiology. This has left a gap in our knowledge, and unfortunately this gap has made study of the primary events associated with information processing impossible. To try to fill this gap by purely mathematical treatment of a presumed diffuse nerve net will lead only to the development of hypothetical concepts, the correctness of which cannot be tested.

2. A New Approach

Our attempts to understand the nervous system can be likened to an attempt to deduce the function of an electronic circuit when we know only the shape of certain parts of the circuit and the electrical events at certain points in the circuit but not the entire design of the circuit. An electronic engineer certainly can combine these parts to establish electronic circuits with very different functions, but to understand how a given circuit functions he needs a circuit diagram from which he can read directly the way the various parts of the circuit operate.

To obtain a circuit diagram of neural centers we must collect sufficient information about the pattern of connections between the neurons in the integrating centers to be able to construct first a wiring diagram for the centers. We also must know the functional characteristics of the various components, which in the case of nervous centers is simplified by the fact that there are only two possibilities: a neuron exerts either an excitatory or an inhibitory influence. With this combined knowledge, it should be possible to draw a circuit diagram that in a rather direct way describes the possible patterns of neural interactions. With a knowledge of the input to and the output from the center and of the behavior of some types of neurons within the center, it should then be possible to deduce the pattern of interactions most likely to be the basis for a particular type of information processing.

In case a circuit diagram is too complicated to be read directly, a computer simulation can be used in which the values for excitation and inhibition assigned to various neural connections are varied until the output signals and the activities of the neurons accessible for recording are correct. Recordings of these signals fulfill a function similar to that of the recordings from test points in an electronic circuit used by the electronic engineer to locate a malfunction in the circuit. The simulated neural circuit malfunctions as long as the recorded behavior of the neurons is not matched.

The development of the method for making three-dimensional reconstructions from electron micrographs of serial sections (Sjöstrand, 1958) has opened up entirely new possibilities for analyzing the organization of the nervous system. With this technique the processes of the neurons can be observed irrespective of their size because the imaging reveals the plasma membrane of the cells. The entire tree of branching dendrites and axons can be analyzed in this manner. Since all neurons in the analyzed piece of tissue can be studied with the same clear definition and therefore can be analyzed simultaneously in the same region, it is possible to reveal the patterns according to which a population of neurons are connected, which is impossible in an analysis of individual stained neurons. Thus, three-dimensional reconstructions can supply information crucial for the construction of a circuit diagram completely describing the pattern of neural connections within an integrating center. This circuit diagram can then replace any mathematical model of nerve nets with precise information regarding the organization of these nerve nets.

This entirely new approach to the analysis of neural centers means that a very big step can be taken beyond the light microscopic or electron microscopic descriptive analysis of individual neurons made partially visible by metal impregnation. Unfortunately, however, the difference in aim between this new approach and that of Ramon y Cajal necessarily restricted in it's aim by the technical limitations is not generally acknowledged.

3. The Choice of Circuitry

In an exploration of the new approach, it is important to choose a simple circuit. On the other hand, the nerve center must be sufficiently complex that the neurons within it can process information, not just contribute to channeling and summation of information. The input and the output of the circuit must be known, and the excitatory and inhibitory components should have been identified. Signal transmission within the circuit must be precise, with a minimum of background noise due to nonsensical connections between neurons.

The circuitry of the retina appeared to be exceptionally favorable with respect both to simplicity and to application of the electron microscopic technique. Part of the retina's circuitry is very condensed, with very minute neural processes. Basic circuits are therefore confined to small volumes of tissue, which at an early stage of the method's development was technically advantageous. However, obtaining a complete series of very thin sections for three-dimensional analysis of retinal tissue required high precision and a highly developed technique for tracing the neural processes. Since the thickness of the thinnest processes was only 600 Å, the thickness of the sections could not exceed 500 Å, and no single section in the series could be missing without ruining the possibilities of tracing the processes.

This situation is unique to the retina. Neural processes so minute are not likely to be found in other parts of the nervous system. With the current development of the technique, three-dimensional analysis is technically considerably easier to pursue in other integrating neural centers because much longer series of sections can now be produced. Series of 500 sections have been produced more or less routinely in our laboratory.

4. Some Examples of Information Processing in the Retina

As pointed out above, an analysis of this kind requires that the part of the nervous system under investigation be capable of some type of information processing other than simple summation and channeling of signals. The retina qualifies beautifully in this respect and represents in fact a neural center comparable to centers in the brain with respect to its capability to process information. Some types of information processing by the retina are known, but it is highly likely that what is known represents only a small fraction of the information processing that takes place there.

A wide discrepancy exists between the number of photoreceptors that furnish the input into the circuits and the number of channels offered by the ganglion cells that connect the retina to the brain. The organization of the retinal circuitry is therefore characterized by considerable convergence. Full use of the information picked up by the photoreceptors requires that the retina at least partially processes this information and that the information is transmitted to the brain in a coded form.

The ganglion cells are located close to the vitread surface of the retina. They receive their input from bipolar cells that connect the ganglion cells to the photoreceptors. Because of convergence and overlap a large number of photoreceptors distributed over a certain area are connected to each ganglion cell. Light falling on these photoreceptors can evoke a response in the ganglion cell. However, light stimulating photoreceptors outside this area can also activate the ganglion cell indirectly. We can therefore distinguish between a stimulation of the ganglion cell by photoreceptors that are connected directly to the ganglion cell by bipolar cells and a stimulation by photoreceptors that are not connected in this manner to the ganglion cell. The entire area of the retina from which a ganglion cell response can be evoked is referred to as the ganglion cell's *receptive field*, a concept introduced by Hartline (1938, 1940).

The area over which photoreceptors are connected to the ganglion cell by bipolar cells is defined as the center. It measures less than one millimeter in diameter. The area surrounding the center within which photoreceptors indirectly affect the ganglion cell is referred to as the *surround*. It can extend up to three millimeters from the center (Daw, 1968). It increases when the illuminance increases.

The concept of the receptive field also applies to other neurons in the retina and, for instance, the size of the center of the receptive field of bipolar cells reflects the extent of the branching of the dendritic tree of the bipolar cells, with the dendrites being connected to photoreceptors.

Ganglion cells can respond specifically to particular modes of stimulation such as movement, and their response can depend on the direction of the movement. Such ganglion cells were discovered by Barlow and Hill in the rabbit retina in 1963. They are characterized by responding with a burst of spikes when an image of either a bright or a dark object moves in one direction over the center of the receptive field while a movement in

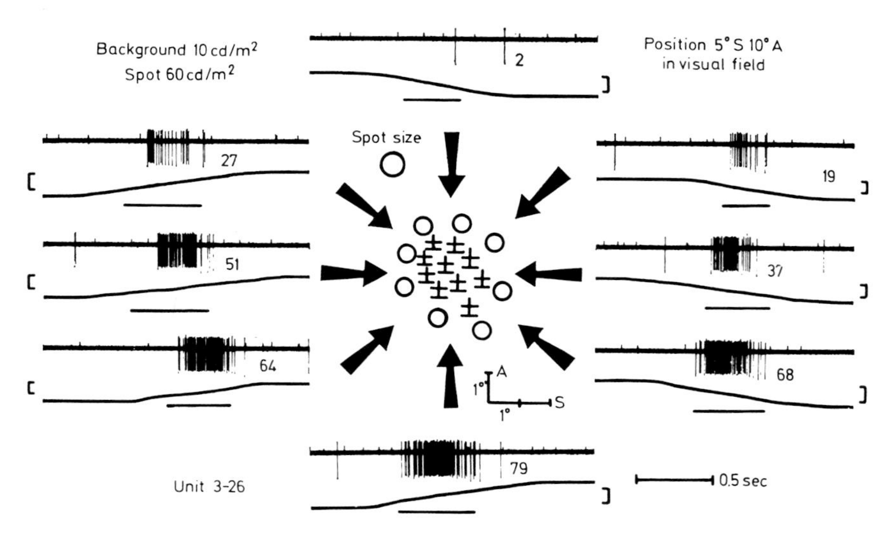

Fig. 1.2. Recordings of directionally selective ganglion cell responses in the rabbit retina to a small spot of light moving across the receptive field in different directions (indicated by arrows). The symbol ± indicates spots at which the flashing of a stationary spot of light caused the ganglion cell to respond to both the turning on and the shutting off of the light, an "on/off" response. The rings indicate spots at which no response could be evoked. The lower of each pair of tracings shows the movement of the spot of light across the receptive field, and the horizontal bar indicates when the spot moved within the center of the receptive field. The upper tracing shows the response of the ganglion cell. The numbers indicate the number of spikes fired. Movement from top to bottom is the null direction, and movement from bottom to top is the preferred direction. The responses to movements become weaker as the direction approaches that of the null direction. From Barlow *et al.* (1964).

the opposite direction does not evoke any response (Fig. 1.2). This response type is one example of an integrated response, and it is remarkable that the retina can carry out the required integration with the few types of neurons that make up its circuitry.

Directionally selective responses have been observed in a variety of species such as the pigeon (Maturana and Frenk, 1963) and the ground squirrel (Michael, 1968), and have been studied in detail by Barlow and co-workers (Barlow and Levick, 1964, 1965; Barlow *et al.*, 1964), Oyster (1968), and by Wyatt and Daw (1975).

The directionally selective response in the rabbit retina can be generated by movements that cover a distance considerably shorter than the distance across the center of the receptive field of a ganglion cell. The minimum distance of movement that can generate a response is less than 50 μm. There must therefore be a number of directionally selective units

Fig. 1.3. The boundary effect. Although the opacity is uniform within each field, the brighter fields appear brighter and the darker fields appear darker at the boundaries.

within the receptive field of the ganglion cell. That directionally selective responses represent an important type of information processing is shown by the fact that up to 40% of the ganglion cells in the rabbit retina respond in this way.

A second example of information processing in the retina is the response by ganglion cells to the introduction of an edge of a dark image in the receptive field (for review see Levick, 1967). Several different types of such "edge detectors" have been observed.

Still another interaction within the retinal network leads to an enhancement of the contrast. The effect is illustrated by the Mach bands discovered in 1865 by Ernst Mach. Contrast enhancement is also involved in the effect observed at the boundary between two fields of different brightness, the *boundary effect* which is a special case of the Mach band phenomenon. We perceive the bright field to be brighter and the dark field to be darker at the boundary than in areas located away from the boundary in spite of the fact that the luminance within each field is uniform (Fig. 1.3).

We can conclude that the retina offers several examples of neural integration of information picked up by the photoreceptors. It therefore qualifies as a nervous center with the capability of information processing.

The rabbit lacks stereoscopic vision because the fields of view of its two eyes are not overlapping sufficiently. Instead, the rabbit enjoys a large, panoramic field of view. In animals with stereoscopic vision, some of the integrative activities of the retina have been taken over by brain centers. Presumably, these centers have become considerably more complex than the corresponding retinal centers because they must be able to retrieve coded information transmitted from the retina, while coding and decoding are unnecessary when the integration occurs in the retina.

5. Some Basic Features of the Structure of the Retina

The circuitry of the first information processing center we analyze should be structurally simple, with the different types of neurons and the total number of neurons being as few as possible. The retina appeared to meet these criteria because it contains only five types of neurons, including the photoreceptors.

The retina is connected to the brain through the ganglion cells, the axons of which form the optic nerve. Between the photoreceptors and the ganglion cells there is only one type of neuron interposed, the bipolar cell. Because this chain of neurons connecting the photoreceptors to the brain is so simple, the processing of information must depend heavily on neurons that transmit signals laterally within the retina.

The different types of neurons in the retina were originally identified by Ramon y Cajal (1892) in Golgi-stained preparations, and his observations were later extended by Polyak (1941). During the last 30 years most of their observations have been confirmed following a renewed interest in the study of Golgi preparations. Since the bipolar cells connect the photoreceptors to the ganglion cells, they make synaptic connections at two levels, at the photoreceptor terminals and at the ganglion cell dendrites. At both these levels one finds laterally extending cells, that is, cells extending perpendicular to the bipolar cells. These cells are the horizontal cells at the level of bipolar cell–photoreceptor connections and the amacrine cells at the level of the bipolar cell–ganglion cell connections. This organization is illustrated in a highly schematic way in Fig. 1.4.

The cell bodies of the bipolar cells, horizontal cells, and amacrine cells are located between these two levels of lateral connections. The neural network is therefore divided into two zones, the outer plexiform layer just vitread of the photoreceptor terminals, and the inner plexiform layer at the level of the ganglion cell dendrites. The layer of cell bodies in between is called the inner nuclear layer while the outer nuclear layer refers to the location of the cell nuclei of the photoreceptors. This layered structure of the retina is shown in Fig. 1.5.

Ramon y Cajal and Polyak identified one neuron that they believed to transmit signals in a centrifugal direction from the inner to the outer

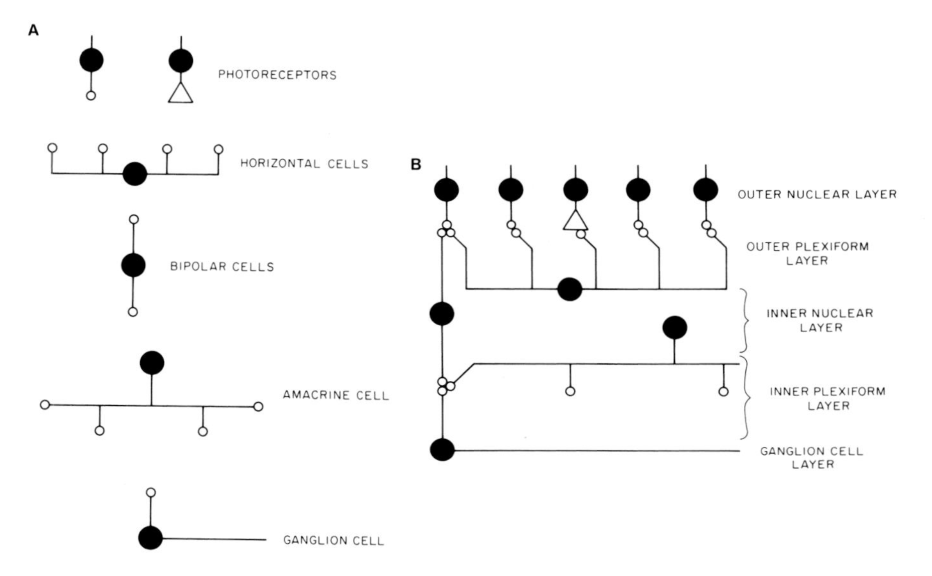

Fig. 1.4. Schematic drawings of the type of neurons in the retina (A) and the way they are connected (B).

plexiform layers. These neurons were called centrifugal bipolars. They have now been renamed "interplexiform cells" (Gallego, 1971). Various types of bipolar cells and horizontal cells have been identified on the basis of the morphology of their dendritic branches and on differences in size.

A nonneural cell type, the Müller's cell, fulfills a function in the retina similar to that of the glia cells in the brain. These cells extend across the retina from its vitread surface to the base of the inner segment of the photoreceptors and contribute processes that are interposed between the neurons.

Of the two plexiform layers, the inner is thicker than the outer. For this reason, it has been assumed that information processing takes place in the inner plexiform layer while summation and transmission of signals take place in the outer plexiform layer.

The circuitry that was analyzed was a part of the outer plexiform layer (Fig. 1.5) and, from the description given above, it is obvious that the circuitry in this layer is about the simplest possible in terms of the number of types of neurons. Only three types of neurons constitute the circuits, the photoreceptors, the bipolar cells, and the horizontal cells (Fig. 1.6). This part of the retina therefore satisfies the requirement for structural simplicity.

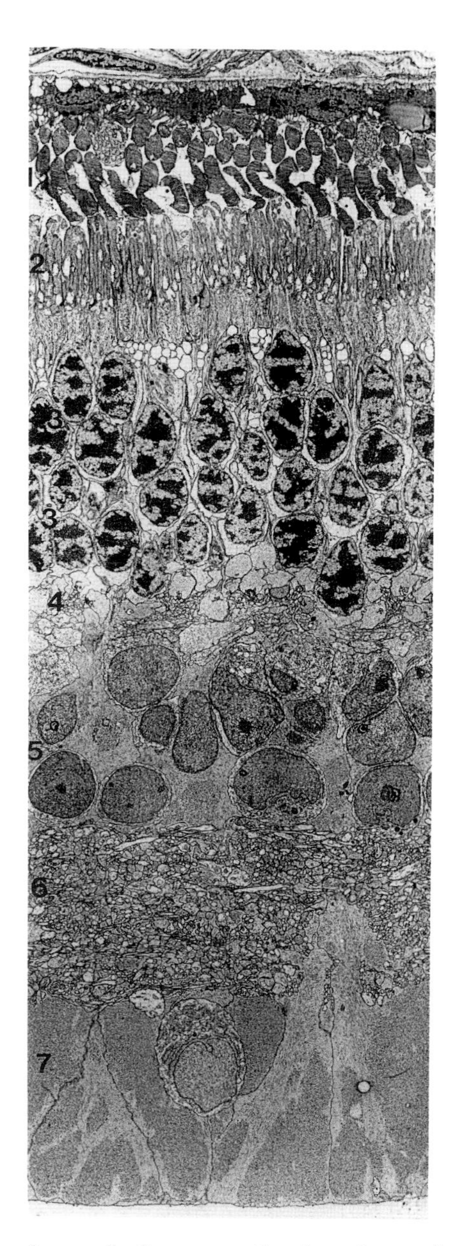

Fig. 1.5. Electron micrograph of a cross section through part of a rabbit retina showing the layered structure as well as the difference in thickness of the outer and inner plexiform layers. 1, Photoreceptor outer segments; 2, inner segments; 3, outer nuclear layer; 4, layer of photoreceptor terminals and outer plexiform layer; 5, inner nuclear layer; 6, inner plexiform layer; 7, ganglion cell layer (one ganglion cell is present within this field of view, illustrating the extensive convergence of the circuitry of the retina in the direction from the photoreceptors to the ganglion cells). 1700×. (Sjöstrand and Nilsson, 1969. Reprinted by permission from "The Rabbit in Eye Research," J. H. Prince, 1969. Courtesy of Charles C Thomas, Publisher, Springfield, Illinois.)

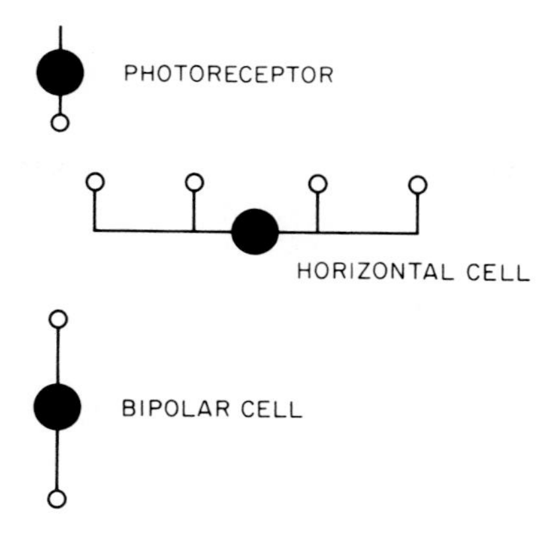

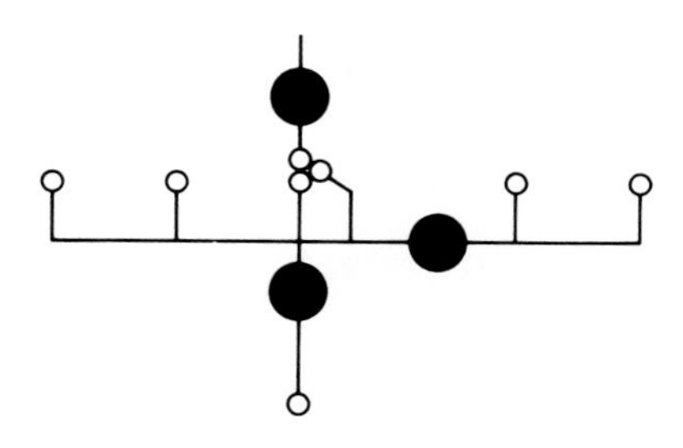

Fig. 1.6. The types of neurons present in the outer plexiform layer and the basic pattern of their mutual connections.

6. A Little History, the Synaptic Vesicles, and the Synaptic Ribbon Complex

The electron microscopic analysis of the photoreceptors pursued in 1952 by applying the improved thin-sectioning technique revealed several special features of the synaptic terminal of the photoreceptors. What are now referred to as synaptic vesicles (Fig. 1.7) were then discovered by Sjöstrand in 1952 as a cytoplasmic component characteristic of a synapse (Sjöstrand, 1953a,b, 1954, 1955, 1956). They were first described as small granules or vesicles, and in the cone terminals of the perch retina they appeared as strongly osmiophilic small granules (Fig. 1.8). Later, de Robertis and Bennett (1954, 1955) described the presence of vesicles in

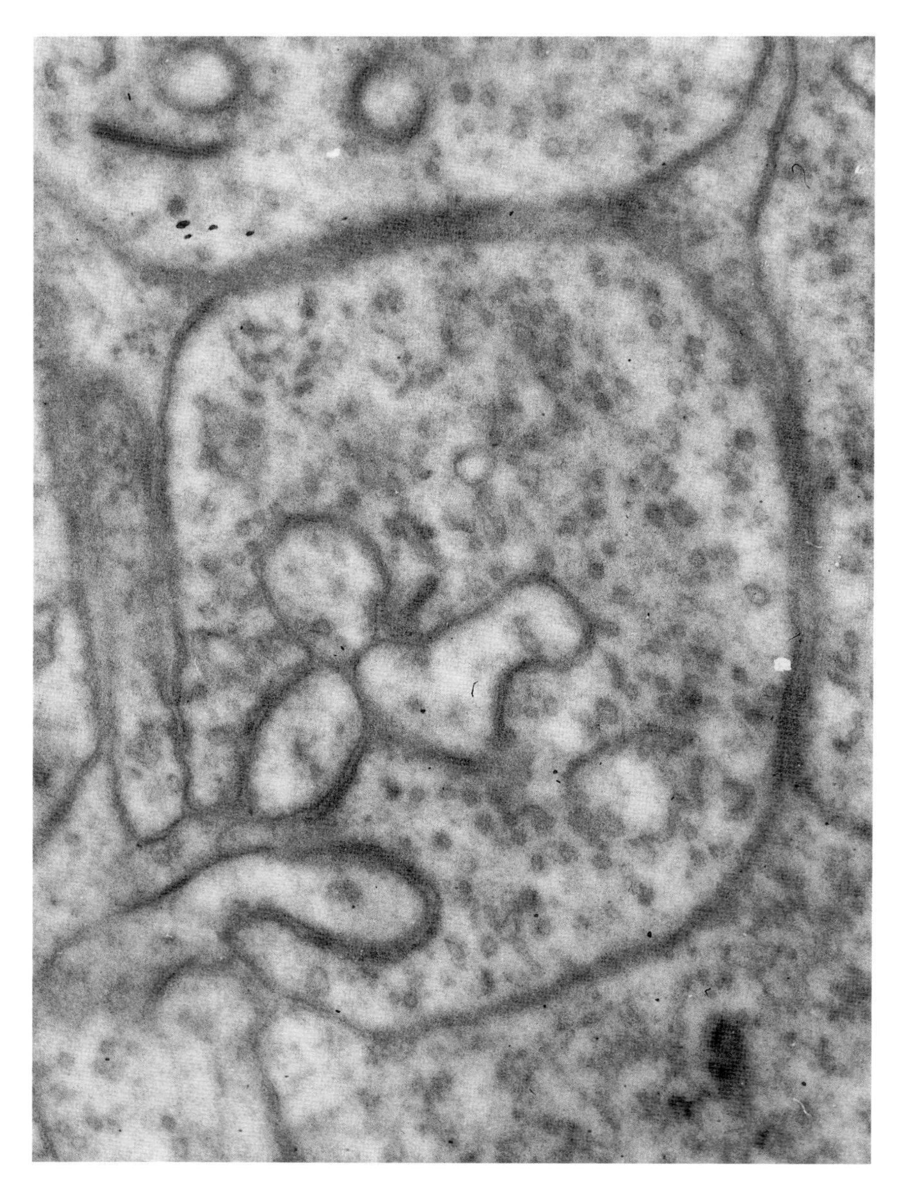

Fig. 1.7. Synaptic vesicles in the rod terminal of the guinea pig retina. This electron micrograph from 1952 clearly shows the synaptic vesicles as a basic structural component of the rod terminal. Also shown is a synaptic ribbon complex, with synaptic vesicles in the endings at the complex. The relationship between these vesicles and neural transmission appeared obscure at that time because vesicles were present on both sides of the synaptic connections. This picture was shown at the Annual Meeting of the German Electron Microscopy Society in Innsbruck in 1953. From Sjöstrand (1954b).

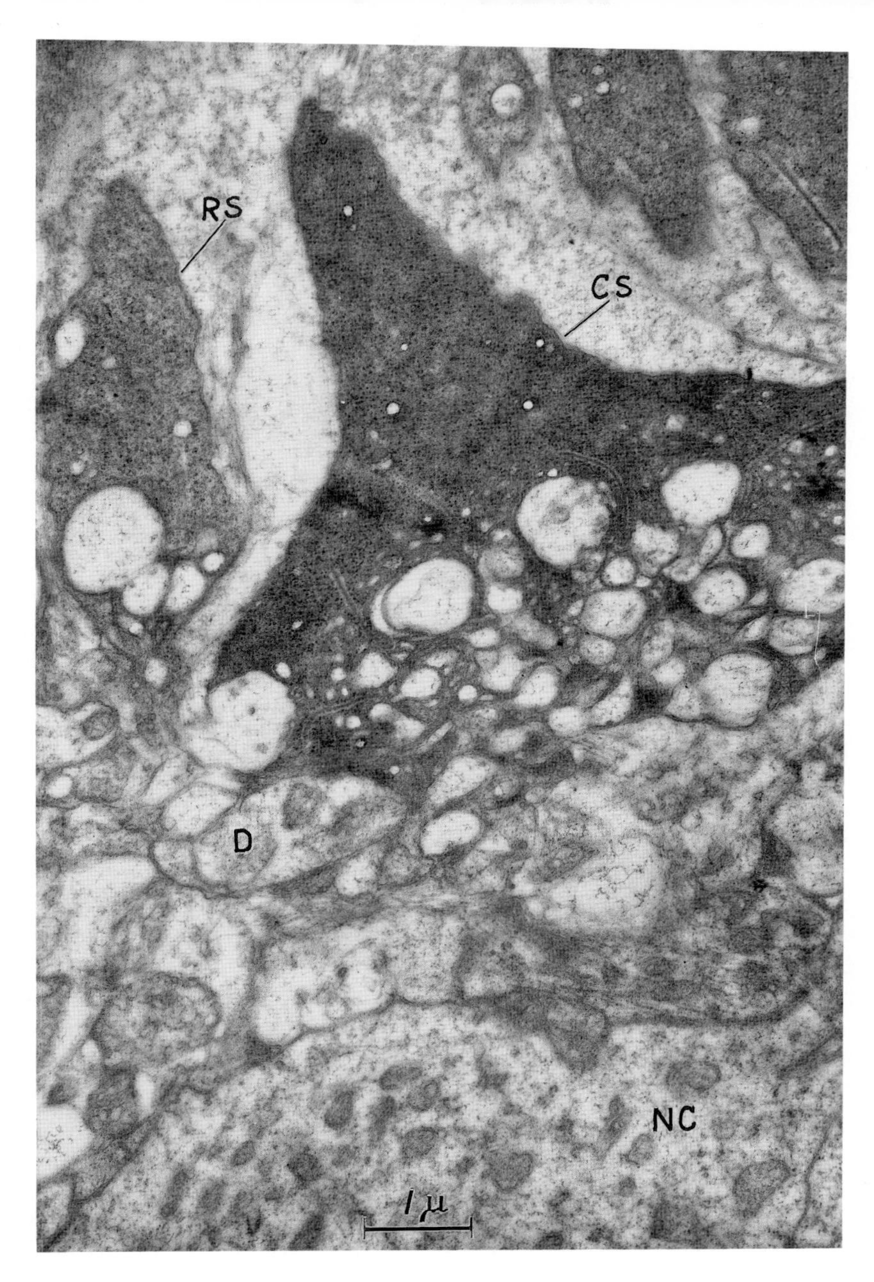

Fig. 1.8. Rod and cone terminals in perch retina. This electron micrograph was taken on January 19, 1952. The synaptic vesicles in the cone terminal to the right (CS) appear as small opaque granules. Such granules are also present in the rod terminal (RS). Such pictures made it unclear at that time whether the vesicular appearance was due to the extraction of material from granules or to real vesicles. Because leakage of material seemed to be a likely effect of the preparatory treatment, the synaptic vesicles were first referred to as granules by Sjöstrand (1953a). 14,000×. From Sjöstrand (1959).

sympathetic ganglia. At that time the identification of synapses in electron micrographs was a problem, a problem that could be solved by analysis of the photoreceptors because their terminal is definitely a synaptic structure.

The analysis of random sections through the retina showed what appeared to be a structurally well-defined complex at the synaptic terminal of the photoreceptors (Sjöstrand, 1953a,b, 1954). A group of vacuoles was located in an invagination of the photoreceptor membrane and was associated with a densely stained structure in the cytoplasm of the terminal (Fig. 1.7).

This complex was first analyzed by comparison of its appearance in sections through a large number of photoreceptor terminals and by deduction of its three-dimensional structure on the basis of a statistical evaluation of the patterns observed in these sections (Sjöstrand, 1954). This evaluation led to certain conclusions regarding the basic topography of the structural components (Fig. 1.9). The opaque structure was found to have the shape of a ribbon and was later referred to as the synaptic ribbon. The vacuoles were found to form two pairs, one of which is located in the deepest part of the invagination. A septum containing the synaptic ribbon extends between the latter two vacuoles and separates two branches of the invagination. The other pair was found to be located closer to the vitread pole of the terminal. The complex is referred to as *the synaptic ribbon complex*.

This study did not allow any conclusions about the nature of the vacuoles to be drawn, although it was clear that the pair of vacuoles located closest to the vitread surface of the terminal were endings of processes located in the outer plexiform layer, as illustrated in Fig. 1.10. Since the nature of these endings could not be established, they were referred to as "vacuoles," which described their appearance in sections. The two pairs of vacuoles were distinguished as the proximal and the distal vacuoles, with the proximal vacuoles being those closest to the center of the photoreceptor. This terminology represents an application of that used in gross anatomy to the cellular level.

Evaluation of the three-dimensional structure of the synaptic ribbon complex on the basis of a statistical analysis allowed only limited conclusions, and it became obvious that only a three-dimensional reconstruction of the complex from electron micrographs of serial sections could furnish further information.

Gay and Anderson (1954) attacked the problem of three-dimensional reconstruction based on electron micrographs of mitochondria, but they made no large-scale reconstructions. They stumbled on one very simple but crucial detail in the development of the methodology, finding a proper

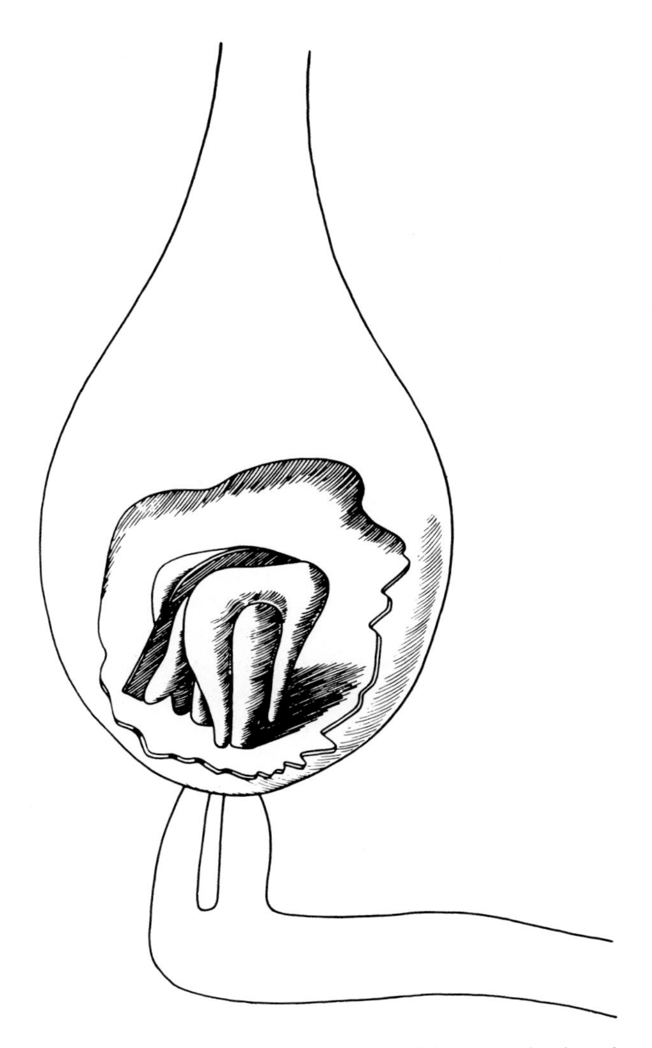

Fig. 1.9. Three-dimensional model of the synaptic ribbon complex based on statistical evaluation of a large number of cross sections through the rod terminals of guinea pig retina. From Sjöstrand (1954a).

support for the ribbon of serial sections that did not obscure any parts of the sections. This problem was solved by the observation that the sections could be mounted on a Formvar film of regular thickness extending over a circular hole with a diameter of 1.5 mm, provided that the beam current in the RCA EMU 1 electron microscope was reduced by changing the gun bias (Sjöstrand, 1958). The electron microscopes in those days were equipped with a minimum of controls, and in order to reduce the beam current it was necessary to unscrew the Wehnelt cylinder (cathode cap).

The reduction of the beam current was important when the series of sections mounted on the one-hole specimen supports were photographed because with a single condenser lens the films broke when examined with the beam current at a normal value. To increase the thickness of the support film would have reduced the contrast, which was low because the sections had not been stained.

The series of sections that were used for the first three-dimensional reconstruction of a photoreceptor terminal (Sjöstrand, 1958) had not been stained because the material was prepared before the introduction of a useful technique for section staining. It was therefore necessary to obtain maximum contrast by preparing very thin sections, only 200 to 300 Å thick, requiring a microtome of exceptional performance for that time that allowed such thin sections to be cut reproducibly and routinely (Sjöstrand, 1953c). The average thickness of the sections used for the first three-dimensional reconstruction of a photoreceptor terminal in 1958 was only 250 Å. This thickness was confirmed by the shapes of neural components in the model of the terminal that was built on the basis of this thickness of the sections. The model showed that the shapes of the neural processes were not distorted in the third dimension.

The section staining technique has radically changed this situation. Excellent contrast can now be obtained in electron micrographs of sections that are 5 to 10 times thicker. It is now the dimensions of the neural processes that determine the maximum allowable thickness of the sections. As a rule of thumb, to guarantee that the neural processes appear in the sections as cut profiles, the thickness of the sections should be less than the diameter of the thinnest processes.

The low contrast of the electron micrographs used in the first reconstruction made the tracing of the neural processes very difficult. It was particularly difficult because rigorous criteria were applied to determine continuity between profiles of neural processes in consecutive sections. This explains why no continuity between the two proximal vacuoles and any process in the outer plexiform layer could be revealed with certainty. With less rigorous requirements, it might have been possible to suggest such continuities.

Figures 1.10 and 1.11 show two sections of the series of sections used for this three-dimensional reconstruction. The contrast in these pictures has been enhanced by triple printing. Figure 1.12 shows the model built when the first photoreceptor terminal was reconstructed (Sjöstrand,

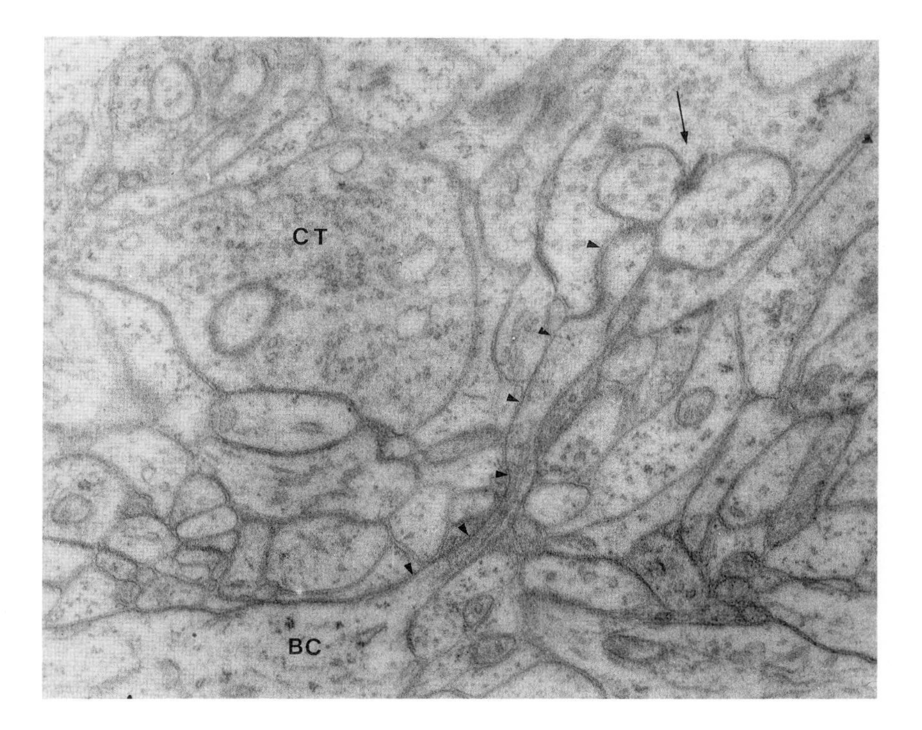

Fig. 1.10 and 1.11. Electron micrographs from the series of sections used for the first three-dimensional reconstruction of a photoreceptor terminal. From Sjöstrand (1958).

Fig. 1.10. A bipolar cell process (arrowheads), ending at the synaptic ribbon complex (arrow). This process was traced to the soma of a bipolar cell, part of which is shown in the lower left corner (BC). To the left is a cone terminal (CT). Because the specimens for the reconstruction were not section stained, only very thin sections could be used to achieve a reasonable contrast. The average section thickness was 250 Å. Triple printing was used to obtain the enhanced contrast shown. Unfortunately, this contrast-enhancing technique does not improve the conditions for tracing the processes. 33,000×.

1958). It clearly confirmed the topographic relations that had been deduced on the basis of the statistical analysis. The distal vacuoles were found to extend into the outer plexiform layer, and one distal vacuole was traced to the soma of a bipolar cell (Fig. 1.10). It was assumed that both distal vacuoles were endings of bipolar cells, and the vacuoles' extension into the outer plexiform layer indicated that the two distal vacuoles were the endings of different bipolar cells.

The proximal vacuoles could not be traced further than to the vitread pole of the terminal because they narrowed considerably as they approached the photoreceptor membrane. No clear communication with

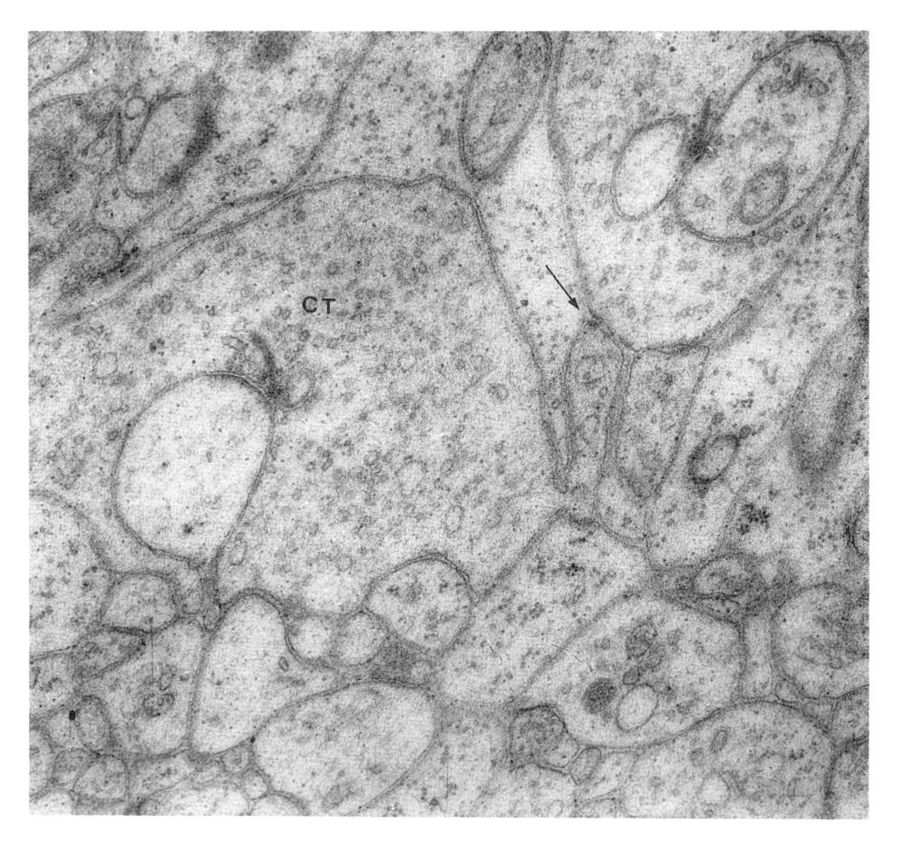

Fig. 1.11. A photoreceptor process from a cone (CT) contacts the vitread pole of the rod terminal (arrow) shown in Fig. 1.10. The rod terminal was connected to several cone terminals. The cone process was shown in other sections to mix with other processes at the vitread pole of the rod terminal, including one process identified as a bipolar cell end process. From Sjöstrand (1958).

any structure outside the terminal could be established because the boundaries of the vacuoles became fuzzy as a result of the low contrast.

Several rod terminals in the guinea pig retina were later reconstructed, and a certain variation in the structure of the synaptic ribbon complex was observed (Sjöstrand, 1965). The number of neural end processes varied from two to four. In these reconstructions all vacuoles were identified as end branches of neural processes located in the outer plexiform layer. In Golgi-stained specimens analyzed with the electron microscope, Stell (1965) showed the proximal vacuoles to be end branches from horizontal cells. He also confirmed that the distal vacuoles were bipolar cell endings.

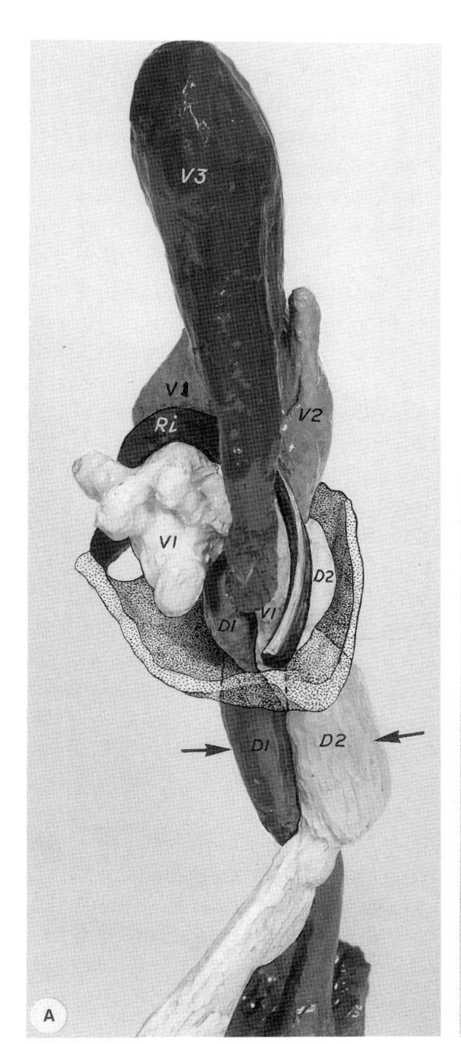

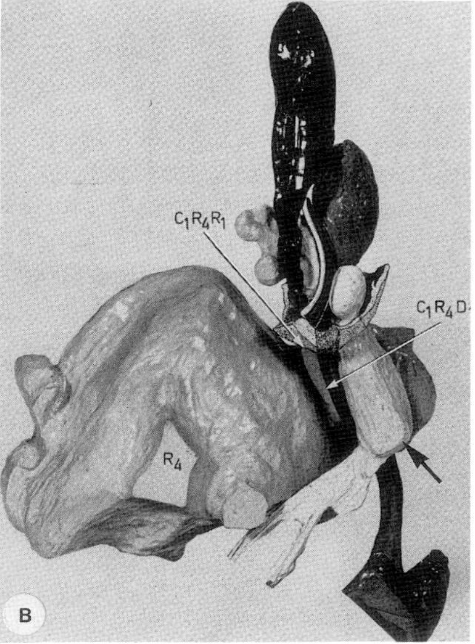

Fig. 1.12. (A) Three-dimensional reconstruction of a rod terminal. V1 and V2 are the proximal vacuoles that extend far in a sclerad direction (up) beyond the distal vacuoles, D1 and D2. Ri is a synaptic ribbon. D2 was traced to the soma of a bipolar cell. (B) Connections between a cone terminal, F4, and the reconstructed rod terminal. The cone connections also involved the bipolar cell end branch at the vitread pole of the rod terminal. C_1,R_4,R_1 indicates connection 1 of receptor 4 (cone) to rod 1, and C_1,R_4,D_1 indicates connection 1 of receptor 4 (cone) to bipolar cell dendrite 1. The widened parts of the bipolar cell endings at the vitread pole of the rod terminal (arrows), which provide large areas for contacts with cone processes, are clearly shown. It was thus established that cone processes made contacts both with the rod terminal and with dendrites of bipolar cells at the vitread pole of the rod terminal. Three cone terminals made connections with this rod terminal and with dendrites just outside the terminal. From Sjöstrand (1958).

Incomplete three-dimensional reconstructions of rod terminals were made by Missotten and co-workers (1963; Missotten, 1965), who confirmed some of the basic features of the synaptic ribbon complex, although they interpreted the topographic relations of the end branches in a different way. They therefore introduced a different terminology and distinguished between centrally located terminal branches that correspond to the distal vacuoles and laterally located branches that correspond to the proximal vacuoles. This terminology is misleading with respect to the overall topographic relationship, and the explanation for this terminology is that Missotten and co-workers did not reconstruct the most sclerad part of the synaptic ribbon complexes, presumably because it was missing in their series of sections. They were, therefore, unable to observe that most of the laterally located terminal branches extended far sclerad after passing the "centrally" located terminal branches. In random sections, the large size and the width of the terminal branches extending deepest into the terminal also contribute to the impression that they are located laterally.

These observations by Sjöstrand and by Missotten and co-workers contributed only fragmentary information on retinal circuitry and did not allow any correlation between this circuitry and the information processing activity of the retina. Such analysis required considerably more extensive three-dimensional reconstructions.

It is generally accepted that information processing in the retina takes place in the inner plexiform layer because through the light microscope this is the only layer that appears to contain a sufficient number of neural connections to accommodate intricate patterns. Rodieck (1973) clearly has expressed this view the following way: "If the effects of the horizontal cells are ignored for the moment, then the function of the outer plexiform layer reduces to one of summation. Summation of signals from a number of photoreceptors onto a smaller number of bipolar cells naturally results in a potential loss of spatial information across the outer plexiform layer." It appeared important, however, that the outer plexiform layer be analyzed to elucidate the input that the inner plexiform layer receives. The analysis not only required work with a considerably larger volume of retinal tissue than had been analyzed in earlier studies by Sjöstrand (1958, 1965) and by Missotten (1965), but the tracing of the neural processes had to be pursued with considerably higher precision.

7. The Methodology

In a random section through the outer plexiform layer, the neural processes appear as illustrated in Fig. 1.13. This figure clearly reveals the

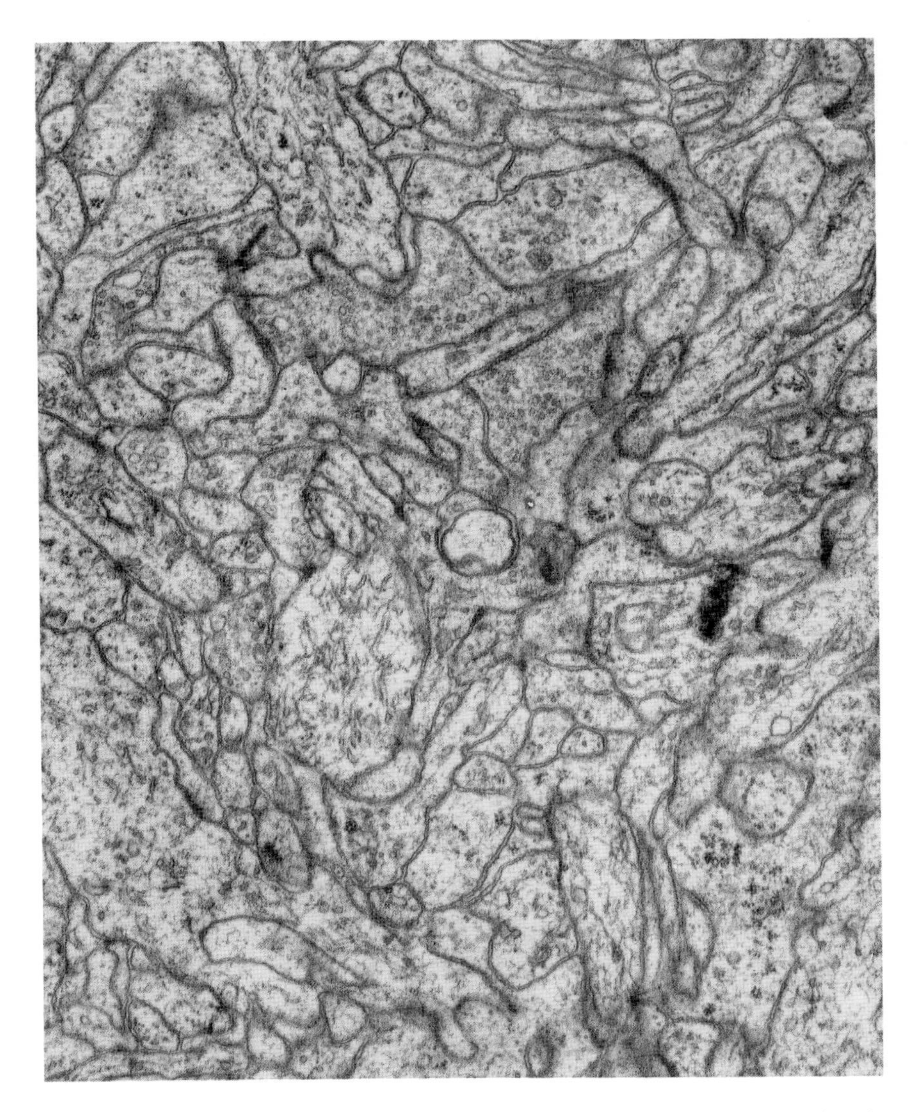

Fig. 1.13. Section through the outer plexiform layer of a rabbit retina revealing the complex arrangement of profiles of neural processes. The picture shows an area at terminal 2, the second cone terminal reconstructed. 19,000×.

compact accumulation of neural profiles. Their arrangement conveys the impression of complexity and randomness. The picture is definitely unintelligible with respect to neural relationships.

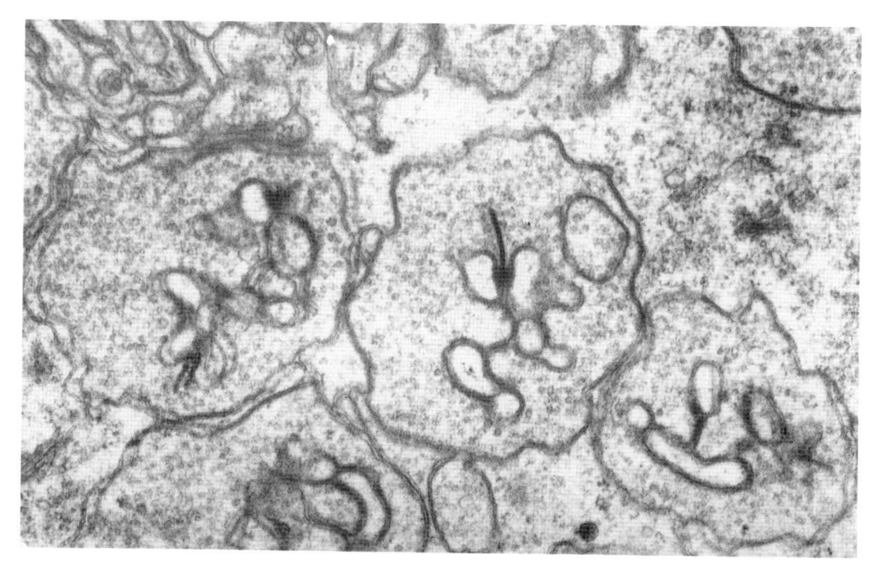

Fig. 1.14. Pictures of two consecutive sections printed on transparent film that have been superimposed to show how profiles in consecutive sections can be traced with high precision when prints on transparent film are used. The fit of the profiles in the two electron micrographs is excellent for the rod terminal close to the center of the picture, but poor for the rod terminal to the left. This discrepancy results from the variation in the degree to which the sections are "compressed' during sectioning. As a consequence of this distortion it is necessary to reposition the micrographs relative to each other when the processes are traced within an area larger than that covered by the profiles of the rod terminal. It is then important to use all possible structures in the pictures as guides when superpositioning the pictures, as can be done with electron micrographs printed on transparent film.

In a three-dimensional reconstruction, the pictures of adjacent sections are superimposed to reveal what profiles in the two sections belong to the same neural process. By superimposing the pictures of adjacent sections in sequence, one can trace the path along which the various processes extend in three dimensions.

This tracing by sequential superpositioning of pictures of adjacent sections requires that the positioning of the pictures relative to each other be highly precise: the thinner the processes and the denser their arrangement, the greater precision is required. Since the correctness of the three-dimensional reconstruction depends on the correct superpositioning of the pictures, this is the most crucial step in the method.

By using positive prints of the electron micrographs on transparent film (Sjöstrand, 1974) as shown in Fig. 1.14, the precision of the tracing of the processes was improved greatly over the earlier technique that involved making drawings of the profiles on sheets of transparent plastic (Sjöstrand, 1958; Missotten, 1965). With the new procedure, all structures extending through both sections in a pair of adjacent sections could be used to guide the positioning of the sections relative to each other.

One test of the correctness of the tracing is based on a complete reconstruction of all processes within the analyzed region. An erroneous tracing leads to nonsensical arrangements of processes, such as closed loops, processes with open ends, separate discontinuous fragments of processes, continuities between processes extending from different neurons, and so on. The reconstructions therefore involved a complete tracing of all processes contacting each photoreceptor terminal in addition to processes adjacent to the terminals.

It turned out that the small size of the end branches of the neurons and the compact arrangement of the processes in this part of the retina exposed the three-dimensional reconstruction technique to a most difficult test. The fact that this circuitry could be analyzed with this technique shows that it can be applied to any other part of the nervous system.

The information was stored in the form of a three-dimensional physical model. When precise shapes of the processes were important to the analysis, an anatomical model revealing these shapes was built. In cases when the model was needed only to link together the profiles of the processes in a series of electron micrographs, a linear model was sufficient.

Linear models can be built quickly. What takes time is the tracing of the processes through the series of pictures and the evaluation of all contact relationships between the processes. This requires direct analysis of the individual electron micrographs, and inclusion of all the pertinent information in a model is not possible.

It has often been suggested that a computer be used for the three-dimensional reconstruction. The computer can assist by memorizing the overall paths of the processes of individual neurons separately, but it cannot assist in the tracing of the processes. The computer therefore cannot contribute to cutting down the time spent on the part of the three-dimensional reconstruction that is most time-consuming. The enormous complexity of neural connections and their closely packed arrangement would make computer retrieval of information rather complicated and would necessitate a second three-dimensional reconstruction from computer-stored data. A physical linear model is a simpler, less time-consuming way of recording the tracing and makes extraction of the pertinent information easy.

An analysis of the neural contact relations requires that all information contained in the electron micrographs be considered. To feed all this information into a computer is not practical, for instance, for the simple reason that the importance of structural features may not be obvious *a priori* but become evident as the analysis proceeds. Viewing the individual electron micrographs is simpler and more reliable than retrieval of selected information stored by a computer.

8. The Material

The first step in the extended three-dimensional analysis was to decide what species of animal to use. The discovery of the directionally selective ganglion cell responses in the rabbit retina that had been made just at the time the work was planned in 1963 appeared to make the rabbit suitable. Several features of the rabbit retina seemed favorable. No color discriminating function had at that time been demonstrated in this retina, and an enormous discrepancy between the number of photoreceptors and the number of ganglion cells within large areas of the retina seemed to indicate that there would be a limited number of circuits for information processing. The complexity of the neural organization, furthermore, appeared to vary considerably in different parts of the retina.

The analysis was started on the premise that there are structurally well-defined neural circuits associated with specific types of information processing and that the neurons do not form a general neural network, the output of which is varied by a varying input. In the latter case, one particular type of information processing would not have a corresponding structural expression, and identification of patterns of neural connections from which functional relationships could be deduced would not be possible.

The complete lack of knowledge regarding such a basic aspect of the organization of a nervous center meant that the analysis was started with considerable uncertainty whether it would turn out much useful information. The three-dimensional reconstruction of the first cone terminal, the adjacent parts of the outer plexiform layer, and the inner nuclear layer took 5 years, not including time spent earlier on the refinement of the methodology. Several tens of thousands of profiles of neural processes were identified as belonging to 70 neurons located within the same region of the reconstructed specimen.

The work was pursued without any knowledge of the significance of the structural relationships, so the three-dimensional reconstruction could not be influenced by subjectivity.

After the three-dimensional reconstruction was finished, analysis of the model took more than one additional year. During this time one feature after another was discovered in which the arrangement of the neural processes followed well-defined ordered patterns. As the analysis proceeded, more and more order was revealed. This discovery of order in such a complex structure was a most rewarding and fascinating experience. The joy of discovery alone made the more than 6 years of work pay off.

Figure 1.15 shows a section through the outer and the inner plexiform layers in the rabbit retina with the section oriented perpendicular to the plane of the retina. Two types of photoreceptors can be identified on the basis of the shape of their terminals. The rod terminals have a rounded ovoid shape while the cone terminals have a conical shape with the base of the cone facing the outer plexiform layer. There is only one synaptic ribbon complex in the rod terminals, but there are many complexes in the cone terminals.

The earlier three-dimensional reconstructions by Sjöstrand and by Missotten and co-workers involved only rod terminals. The three-dimensional analysis of the rabbit retina involved both rod and cone terminals. Figure 1.16 shows one cone terminal and several rod terminals. At the vitread surface of the cone terminals an accumulation of profiles of thin processes can be seen. This accumulation is more or less extensive depending on what part of the retina is viewed.

The extended analysis in which the new methodology was applied involved in its first phase the complete reconstruction of one cone terminal (terminal 1) with all its neural connections, requiring the three-dimensional tracing of processes from 19 neurons contacting the terminal, 9 horizontal cells, 10 bipolar cells, and processes from several neurons located adjacent to the terminal that did not contact it.

The horizontal cells contributed 24 end branches, and the bipolar cells contributed 36 end branches received in invaginations of the photoreceptor membrane. The total number of such end branches was therefore 60. With this large number of end branches it was possible to reveal that neural connections were arranged systematically according to certain combinations.

The processes from the neurons were also traced to six other cone terminals, establishing that the same type of connection was made by one neural process at several different terminals.

The second phase of the extended analysis was the three-dimensional reconstruction of a second cone terminal (terminal 2) adjacent to terminal 1. In this case, processes from 35 neurons were traced, and 18 of these sent branches to terminal 2. Of these, 10 were horizontal cell processes and 8 were bipolar cell processes.

In addition, several rod terminals were reconstructed. At this stage, several hundred end branches contributed by processes from 70 neurons within the same region of the analyzed retina were analyzed.

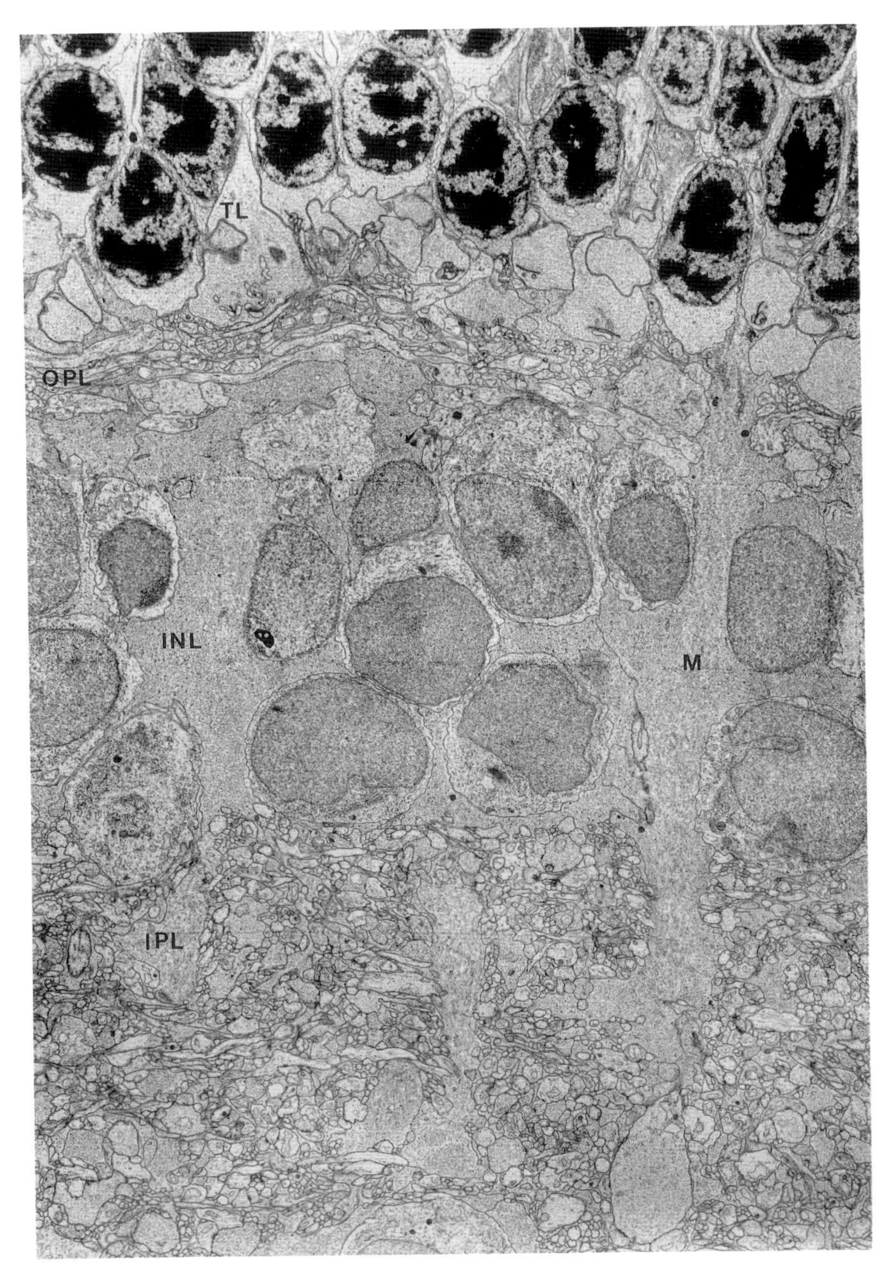

Fig. 1.15. The layers of circuitry in the rabbit retina. The outer plexiform layer (OPL) and the inner plexiform layer (IPL) are separated by the inner nuclear layer (INL) containing the somas of the bipolar cells, the horizontal cells, and the amacrine cells. The photoreceptor terminals are located sclerad to the outer plexiform layer in the terminal layer (TL). Note the staggered arrangement of the terminals, with the rod terminals located sclerad relative to the cone terminals. M, Müller's cell cytoplasm. 4500×.

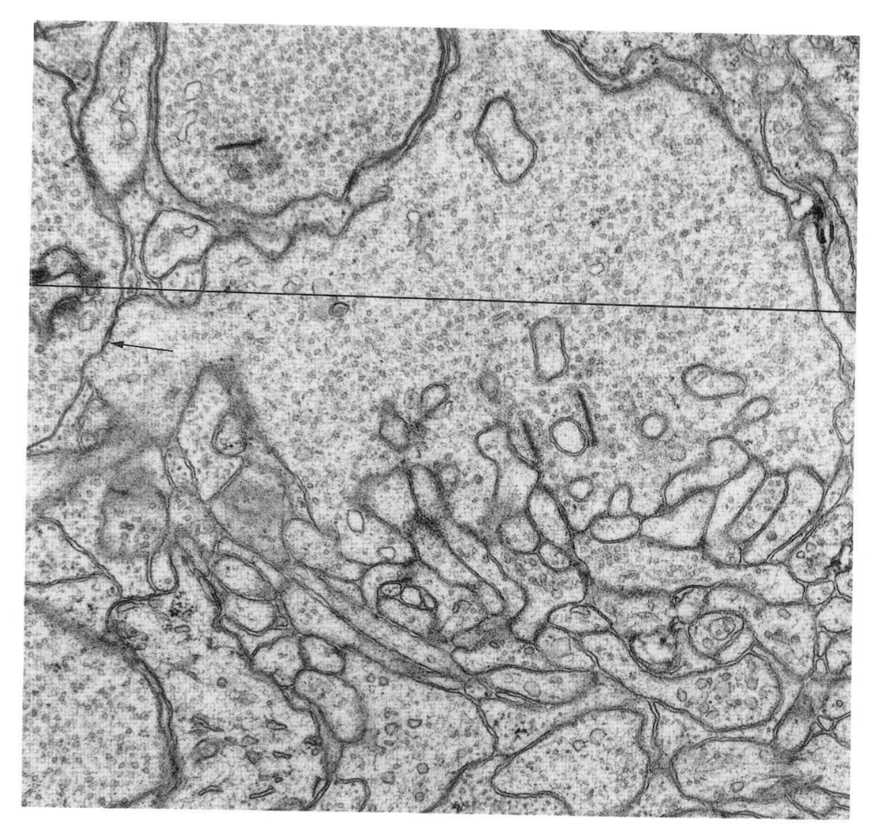

Fig. 1.16. Micrograph of a cone terminal in which the plane of the section is oriented parallel to the long axis of the photoreceptor. At the broad base of the terminal which faces the outer plexiform layer are numerous invaginations of the terminal membrane that receives end branches of neurons. Below the base is a mixture of closely arranged profiles of neural processes. To the left, a connection between the cone terminal and a rod terminal can be seen (arrow). The line across the terminal indicates the orientation of the plane of that sections used for the extended three-dimensional reconstruction. 29,000×. From Sjöstrand (1969).

9. The Three-Dimensional Reconstructions

For greatest efficiency, the plane of the sections was oriented parallel to the plane of the outer plexiform layer. In this way the largest possible volume of this layer was included in the smallest number of sections. A maximum of cross sections through neural processes contacting the terminals are obtained this way, which is favorable because the processes are easiest to trace when cross-sectioned in the case the thickness of the processes approaches that of the sections.

Figures 1.17–1.30 have been chosen from the series of sections used in this analysis to illustrate how the structural features change as we move vitread through terminal 1, through the corresponding part of the outer plexiform layer, and partially through the inner nuclear layer. The plane of the sections is oriented roughly perpendicular to the plane of the section shown in Fig. 1.16.

Figure 1.17 shows cross sections through terminal 1 and through several rod terminals. The irregular shape of the cone terminal is due to the large number of processes extending laterally from the main body of the terminal. Part of one such process is marked by an arrow. While the cytoplasm of the main body of the terminal is filled with synaptic vesicles, part of the lateral process is devoid of such vesicles.

In the main body of the terminal, a number of vacuoles are seen, and a certain systematic trend in the arrangement of the vacuoles can be discerned. At B there is a triad of one small and two large vacuoles. An intensely stained rod-shaped structure is located between the two large vacuoles. This dense structure is the synaptic ribbon, which appears rodlike when the plane of the section is perpendicular to the ribbon. In an adjacent rod terminal, at the upper border of the picture at Z, the ribbon shape is evident because the plane of the ribbon is almost parallel to the plane of the section.

The triad at B is one synaptic ribbon complex out of 16 complexes (indicated by letters) present in this terminal. Certain basic features of the structure of the synaptic ribbon complex are shown here. The complex consisting of three vacuoles is surrounded by the photoreceptor membrane, which forms deep invaginations in which the end branches from horizontal cells and bipolar cells are received. The synaptic ribbon lies in a septum between the two large vacuoles located in two end branches of the invagination. At the edge of this septum is a cytoplasmic density associated with the photoreceptor membrane, as shown at the triad labeled B. Only a narrow space separates the two large vacuoles from the small vacuole, and at higher magnification one can see that the bounding membranes are particularly thin in this region. Synaptic vesicles are present in one of the two large vacuoles. The number of vesicles present in these vacuoles varies considerably, which is explained by variations in the extent to which the vesicles have been preserved.

In Fig. 1.18 we have moved six sections vitread, that is, about 0.3 μm. Six of the sixteen synaptic ribbon complexes are seen in this picture, which illustrates in two pairs of synaptic ribbon complexes, A and F, C and E, that one large vacuole can be associated with two adjacent

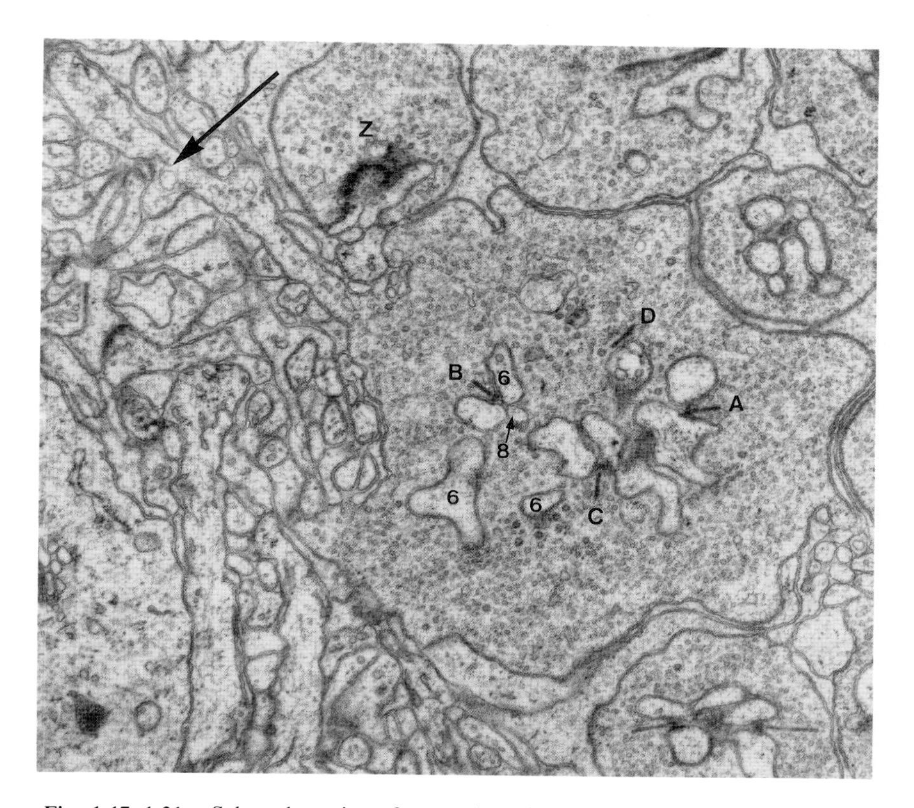

Fig. 1.17–1.31. Selected sections from series of sections through terminal 1. From Sjöstrand (1974).

Fig. 1.17. Section 79. The plane of the section is oriented transversally to the long axis of the photoreceptor, as indicated by the line in Fig. 1.16. Several synaptic ribbon complexes, marked A to D, are present. At complex B the triad arrangement of one synaptic ribbon complex is seen, with two larger profiles of horizontal cell endings and one small profile of one ending of bipolar cell 8. The synaptic ribbon appears as an opaque rodlike structure. Its ribbon shape is revealed at Z, where the synaptic ribbon is oriented almost parallel to the plane of the section. One process from the terminal extends laterally toward the upper left corner of the picture (arrow). 19,000×.

complexes. The triad at synaptic ribbon complex C also illustrates the considerable difference in size between the two large vacuoles and the third small vacuole, which is one ending of bipolar cell 13.

Moving vitread another six sections (Fig. 1.19) reveals the irregular shape of the terminal and a number of processes extending laterally from its main body (arrows). The structure in the interior of the terminal has become more complex, with a large number of profiles partially closely

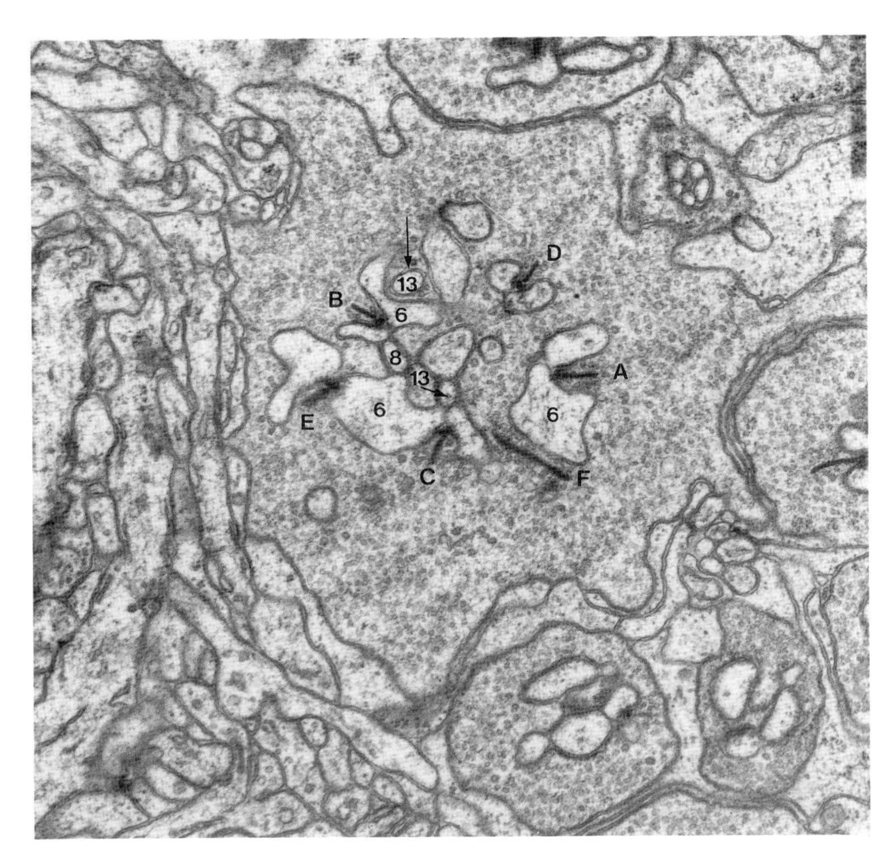

Fig. 1.18. Section 73. The difference in size between horizontal and bipolar cell endings at synaptic ribbon complexes is clearly illustrated. Horizontal cell 6 contributes one large ending that is associated with two adjacent synaptic ribbon complexes, at complexes A and F and at complexes C and E. Bipolar cells 8 and 13 are associated with synaptic ribbon complexes B and C, respectively. In addition, a profile of bipolar cell 13 (arrow) is located in an invagination that is separate from those containing synaptic ribbon complexes. The three-dimensional tracing is required for establishing this relationship. 20,000×.

Fig. 1.19. Section 67. Processes extend in several directions (arrows) from the main body of the terminal. At this level neural profiles are closely apposed in the center of the terminal. This is the most sclerad part of what will be referred to as the subsynaptic neuropil, because these processes are located vitread, "below," the terminal. Bipolar cell 39 (core bipolar cell) appears as a rather large irregularly shaped profile in the subsynaptic neuropil. Bipolar cell 14 contributes two profiles that are not associated with any synaptic ribbon complex. The two endings are located in separate invaginations and they contain numerous synaptic vesicles. Close to the upper right corner is a profile contributed by a process (CP) from an adjacent cone terminal. Between this profile and terminal 1 is an accumulation of profiles that represent part of the subsynaptic neuropil at a rod terminal (arrowheads). Although the cone process contains vesicles, no synaptic vesicles were present along part of the process proximal to the part shown here. 23,000×.

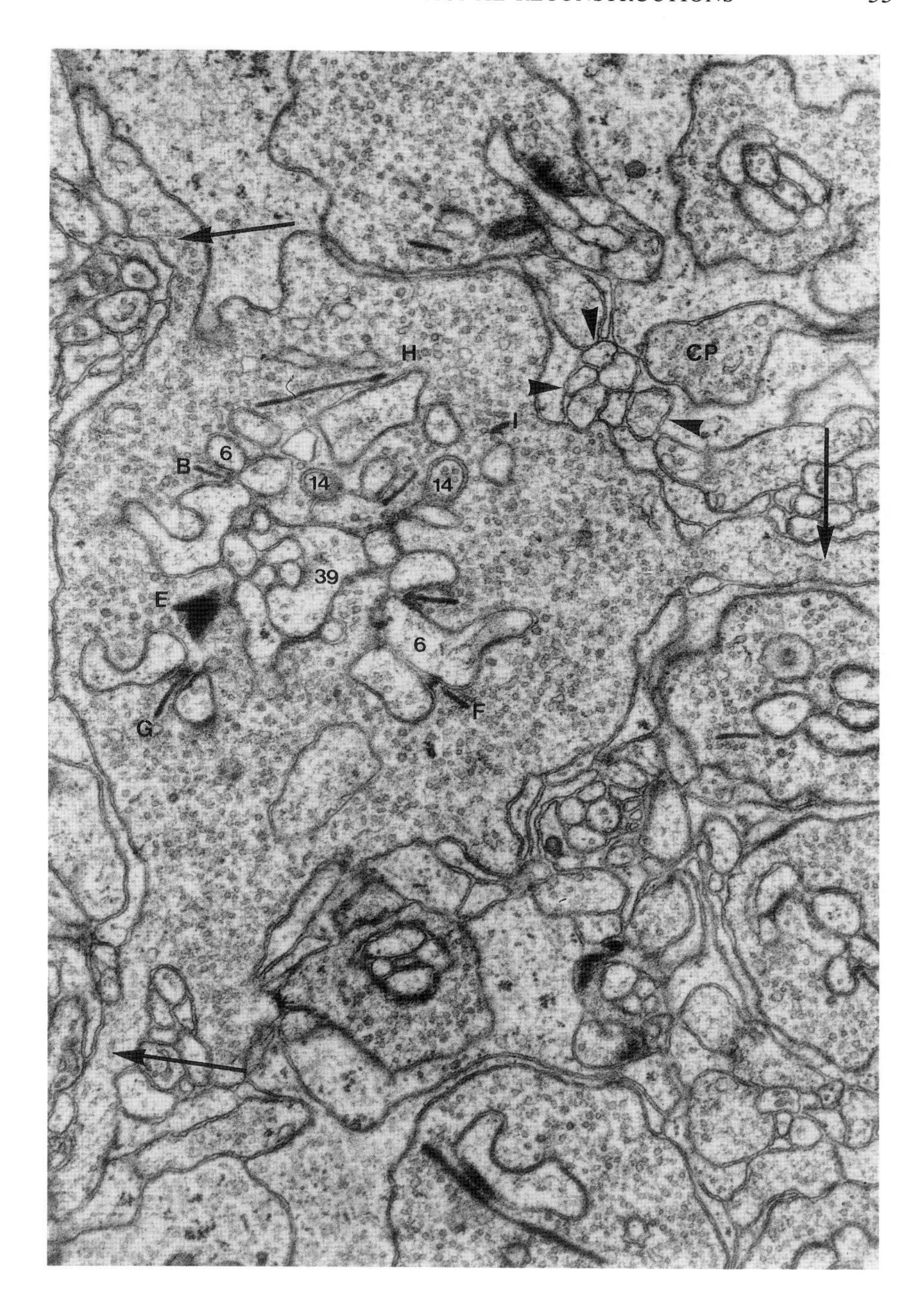
H
CP
I
6
B
14
14
39
E
6
G
F

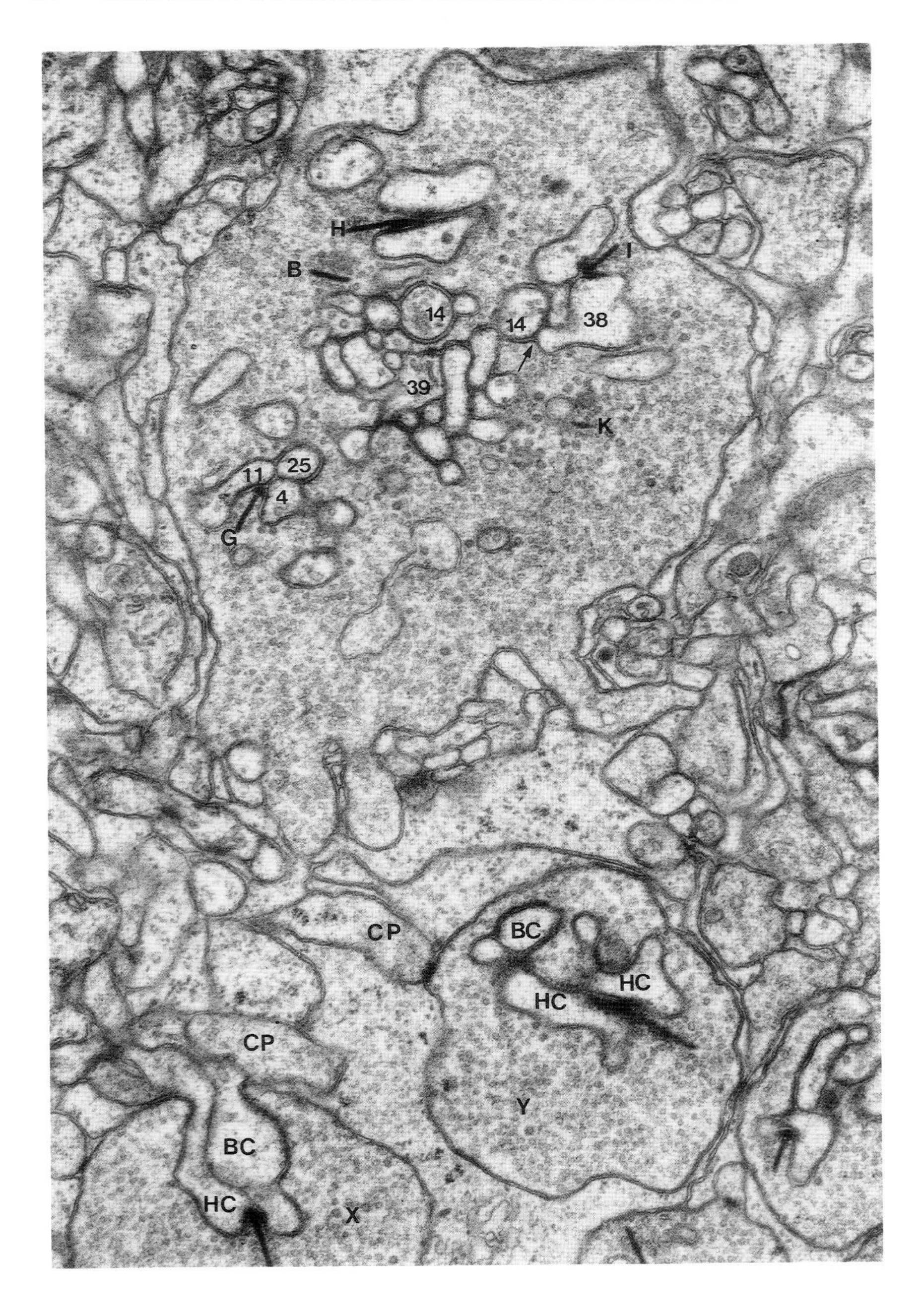
H
B
I
14
14
38
39
K
25
11
4
G
CP
BC
HC
HC
CP
Y
BC
HC
X

packed. In Figs. 1.20–1.26 the structure of the terminal appears progressively more complex as we move further toward its vitread surface. More and more of the neuropil adjacent to the terminal contribute to the picture, while the area occupied by the cytoplasm of the terminal (T) is reduced.

The remaining figures from this series of sections, Figs. 1.27–1.30, reveal the complex structure at levels between the terminals and the inner nuclear layer. Large horizontal cell branches and Müller's cell territories dominate in Fig. 1.30, which shows the 66th section from the one pictured in Fig. 1.17. The distance between these two pictures perpendicular to the plane of the section is about 4 μm.

Figure 1.31 shows the soma of bipolar cell 13, in the inner nuclear layer, one dendrite of which was traced to terminal 1, where it ended with several end branches.

Although the synaptic ribbon complexes can be identified in the pictures and convey an impression of some kind of structural order, only with a three-dimensional reconstruction is interpretation of the pictures in terms of neural circuitry possible.

For the three-dimensional reconstruction, the series of sections was photographed at the lowest suitable magnification to make each picture cover the largest possible area. It was found that a magnification as low as 6400× was appropriate and that at this low magnification the electron micrographs contained all structural information above the level of artifacts. Figures 1.18–1.32 are all prints of electron micrographs used in the reconstruction and the electron optical magnification is only 6,400×. Figure 1.32 shows a higher light optical magnification of one of these electron micrographs.

Fig. 1.20. Section 63. The subsynaptic neuropil has become somewhat more complex and the two endings of bipolar cell 14 have widened. They contain numerous synaptic vesicles, and the apposed neural membranes appear particularly thick and are separated by a space wider than that separating other neural membranes. The end branch of bipolar cell 39 contains synaptic vesicles. A lateral contact between bipolar cell 14 and horizontal cell 38 is seen at the synaptic ribbon complex I (arrow). It is a lateral extension of the endings of horizontal cell 38, which is in contact with bipolar cell 14. This bipolar cell made no triad type of connection at the complex. Bipolar cell 25 contributes an ending to complex G that differs from the usual endings at the complexes. Its contact with the terminal membrane is of the same type as that of bipolar cell endings contacting the terminal outside synaptic ribbon complexes with the apposed membranes being thick and separated by a particularly wide space, while the membrane is thin and the membrane separation is narrow at its contact to the two horizontal cells. The two rod terminals X and Y in the lower part of the picture were reconstructed; Figs. 2.16 and 2.17 show the three-dimensional reconstructions of these terminals. BP, bipolar cells; HC, horizontal cells; CP, cone processes. 23,000×.

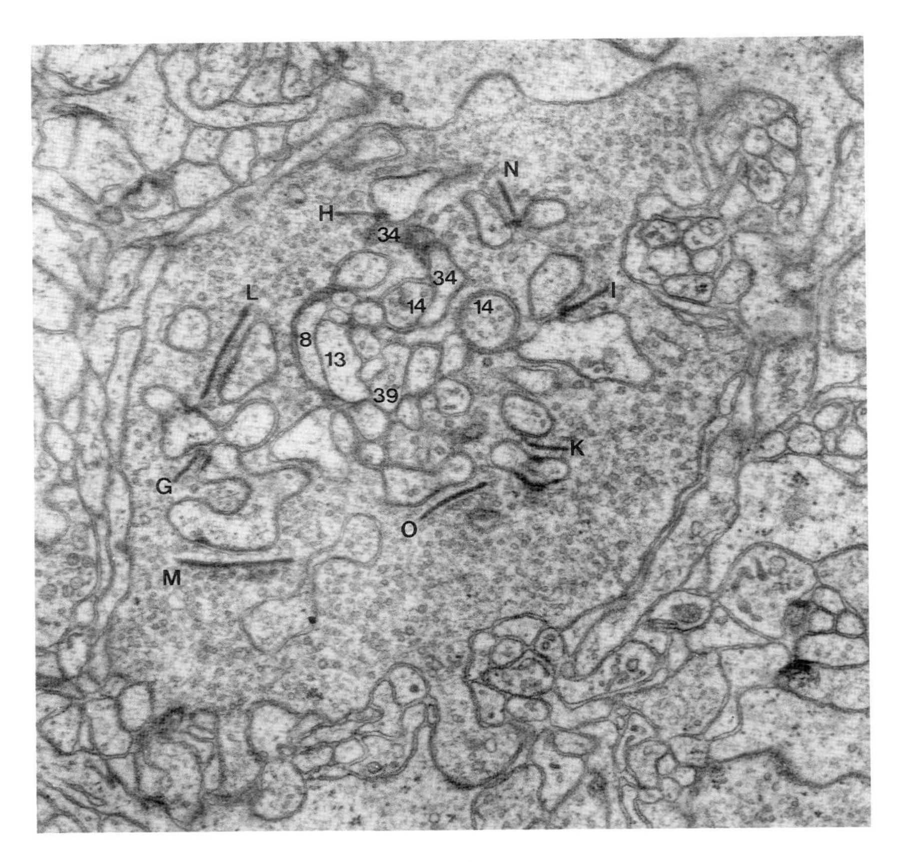

Fig. 1.21. Section 60. At synaptic ribbon complex H, horizontal cell ending 34 extends a rather long branch that ends in a broad contact with bipolar cell 14. Both endings of bipolar cell 14 are thus contacted by extensions from horizontal cell endings at adjacent synaptic ribbon complexes. Bipolar cells 8 and 13 are closely associated, with a particularly close arrangement of the thin neural membranes. 22,000×

It might seem surprising that such a low electron optical magnification is sufficient. However, at 6400× magnification, a membrane 100 Å thick appears as a line 64 μm thick in the electron micrograph. Magnifying the latter by a factor of 4 renders a print in which the membrane appears as a line 0.26 mm wide. At a factor of 10, the line will be 0.64 mm thick, and it is easy to distinguish between membranes of different thickness and to observe differences in the width of the space separating the membranes. The triple-layered appearance of the plasma membranes may be visible,

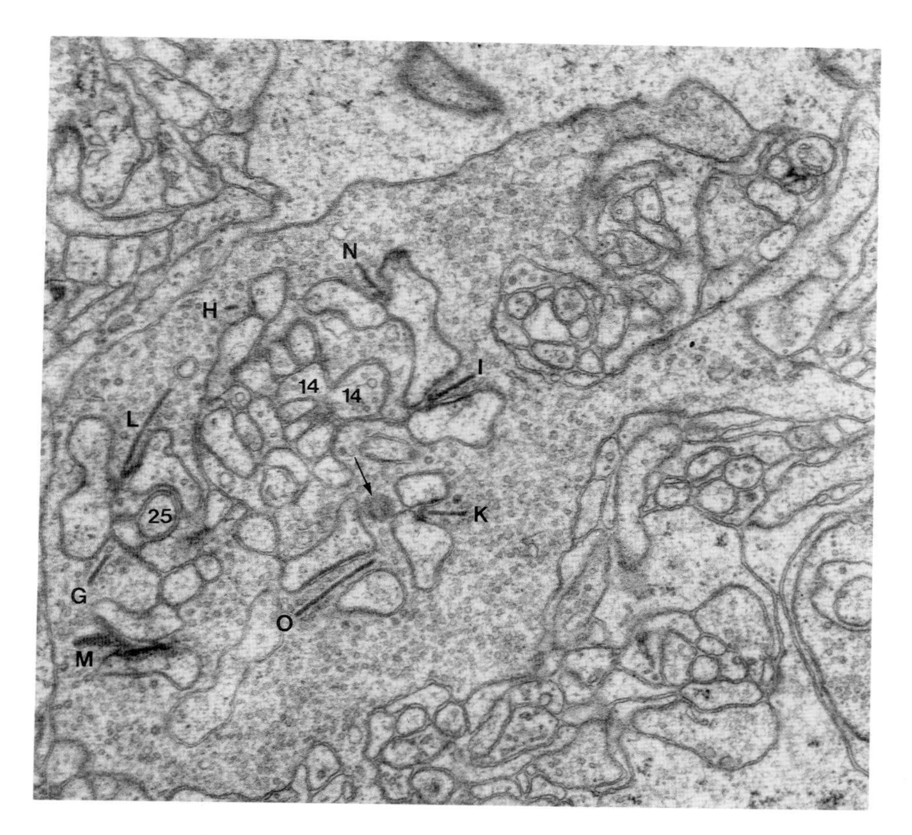

Fig. 1.22. Section 58. The subsynaptic neuropil is more complex at this level. The two endings of bipolar cell 14 that were present in the earlier pictures fuse. At the lower left, bipolar cell 25 contributes an ending that lies in an invagination separate from any synaptic ribbon complex. At complex K the shaded area (arrow) is the sclerad end surface of the same kind of ending from bipolar cell 1, located outside a synaptic ribbon complex. 22,000×.

meaning one can observe structural features even at the level of resolution where artifacts dominate.

To cover an entire section, a number of electron micrographs of each section were produced. Montages made from regular paper prints of these pictures solved the problem of identifying which electron micrographs should be studied as the analysis was extended over a progressively larger area within the sections.

The anatomical model was built from large sheets of transparent plastic. The procedure was as follows: The various profiles in the electron

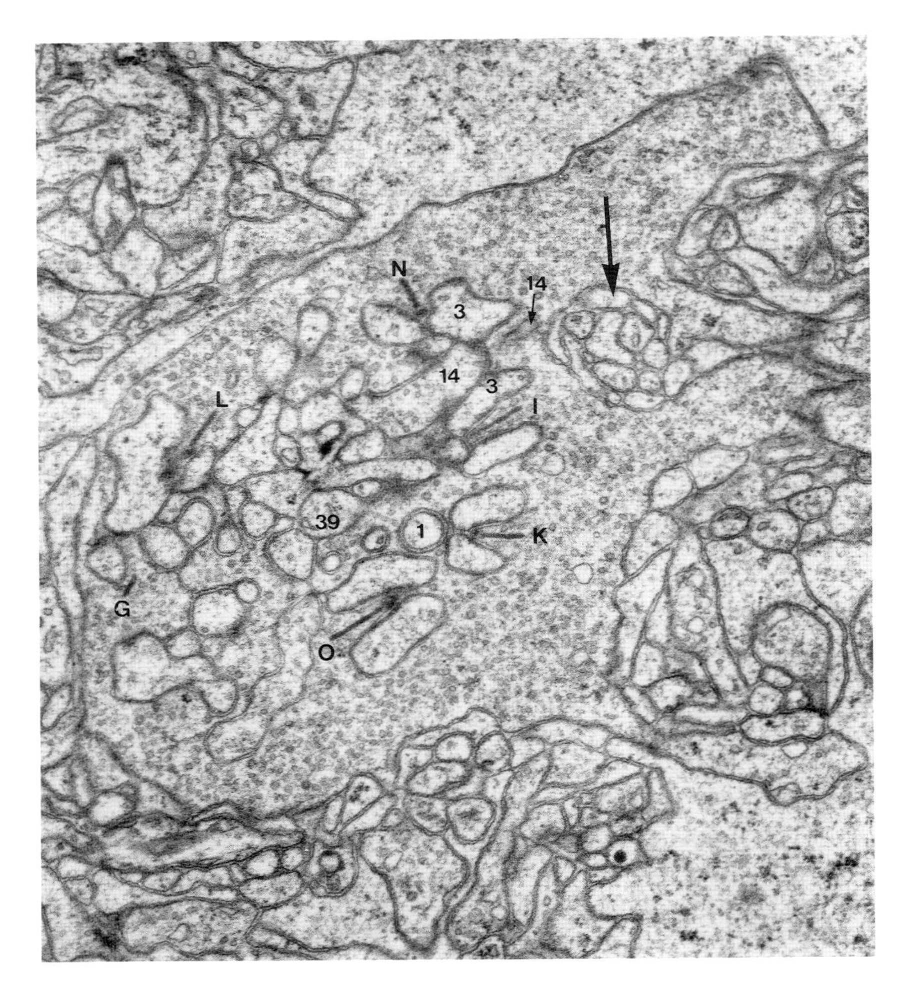

Fig. 1.23. Section 56. A third ending of bipolar cell 14 is cut obliquely and extends between two endings of horizontal cell 3, which it contacts laterally. The ending of bipolar cell 1 close to ribbon complex K is separated from the complex by a thin layer of cytoplasm. The thick membranes with the wide spacing, which are characteristic of bipolar cell endings that contact terminals outside synaptic ribbon complexes, can be seen. This ending is located close to ribbon complex K but does not contact the horizontal cell endings at the complex. Another ending of bipolar cell 1 makes a triad contact at this complex but further in a vitread direction. The neural membrane of the latter ending is not thicker than that of the horizontal cells. Therefore the two types of bipolar cell endings have structurally different membranes. Processes extend laterally from the main body of the terminal, and in the upper right (arrow) two such processes surround the subsynaptic neuropil at a rod terminal located sclerad to the plane of the picture. 23,000×.

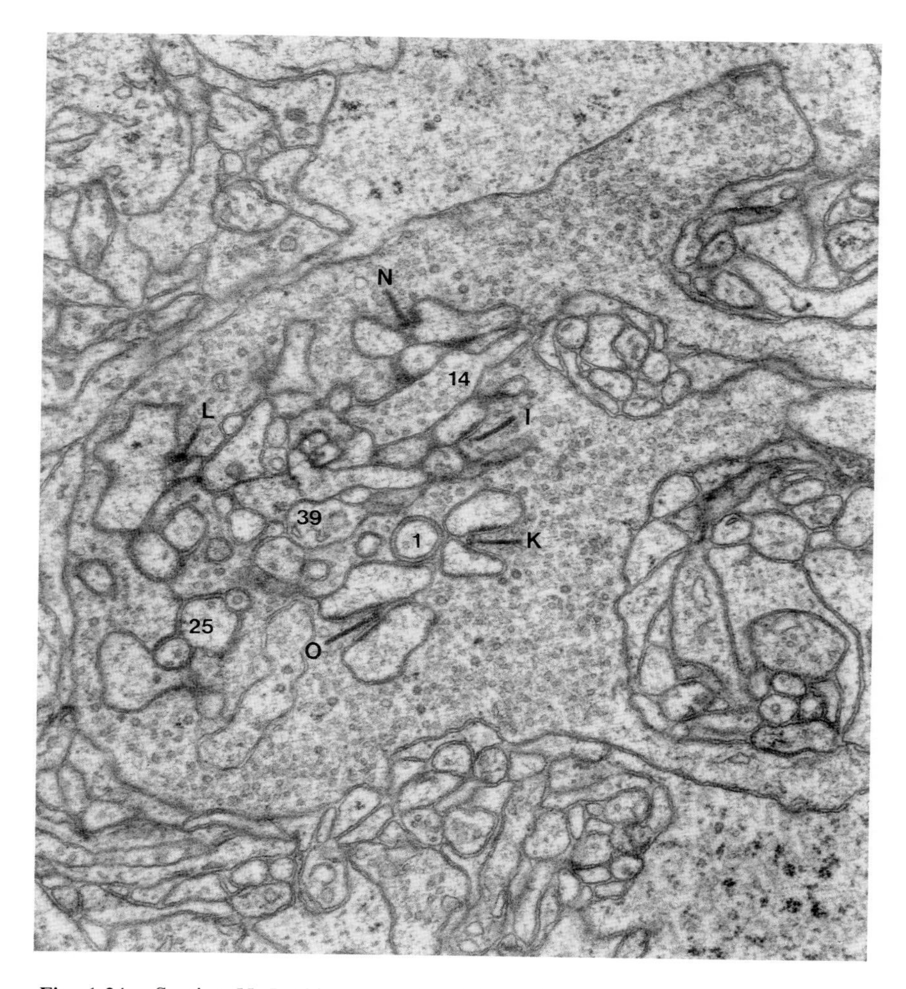

Fig. 1.24. Section 55. In this section that follows that in Fig. 1.23, the third ending of bipolar cell 14 is seen filled with synaptic vesicles, and the ending of bipolar cell 1 close to ribbon complex K is still separated from the complex. In the lower left part of the terminal an ending of bipolar cell 25 contacts the terminal outside the synaptic ribbon complexes. 23,000×.

micrographs that had been printed on transparent film at a total electron and light optical magnification of 26,000× were first traced through the series of sections. In this way the corresponding profiles in adjacent sections were identified. The positive transparent print was then projected onto a large sheet of plastic at a further 10× magnification, and the

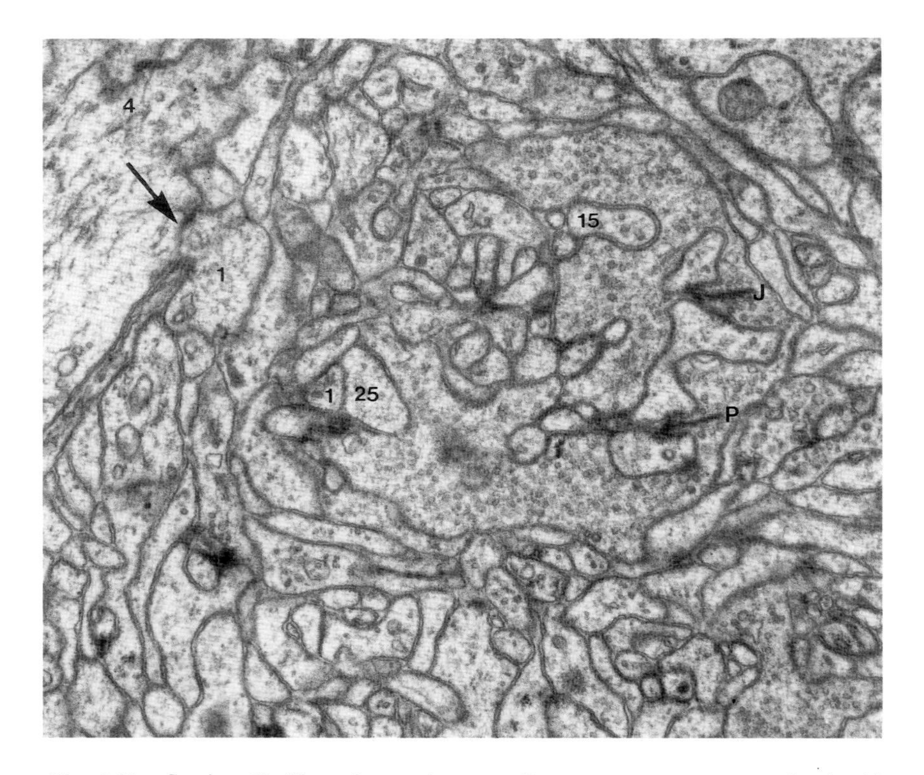

Fig. 1.25. Section 49. The subsynaptic neuropil appears even more complex in this section. Bipolar cell 15 contributes one profile in the upper right part of the terminal. The ending is associated with thick neural membranes separated by a particularly wide space. Bipolar cell 15 does not contact any horizontal cells at this terminal. The arrow at the upper left corner points to a contact between one collateral of bipolar cell 1 and the large horizontal cell 4. The end processes of bipolar cells 1 and 25 are in the same close contact as bipolar cells 8 and 13 shown in Fig. 1.21. 23,000×.

outlines of all profiles were drawn on this sheet. This drawing was thus made at 260,000× magnification.

At the next step, the profiles were cut out with a Rockwell 24-in. scroll saw; thin connections were left between the profiles to preserve their relative positions. The cut sheet of plastic was now positioned on top of the cut out profiles of the previous sections, and corresponding profiles were glued together with a plastic spacer of a size that spaced the plastic sheets at a distance corresponding to a section thickness of 500 Å. Figure 1.33 shows an early stage during the building of the anatomical model, and Fig. 1.34 shows the model in its final stage; a light table is in the

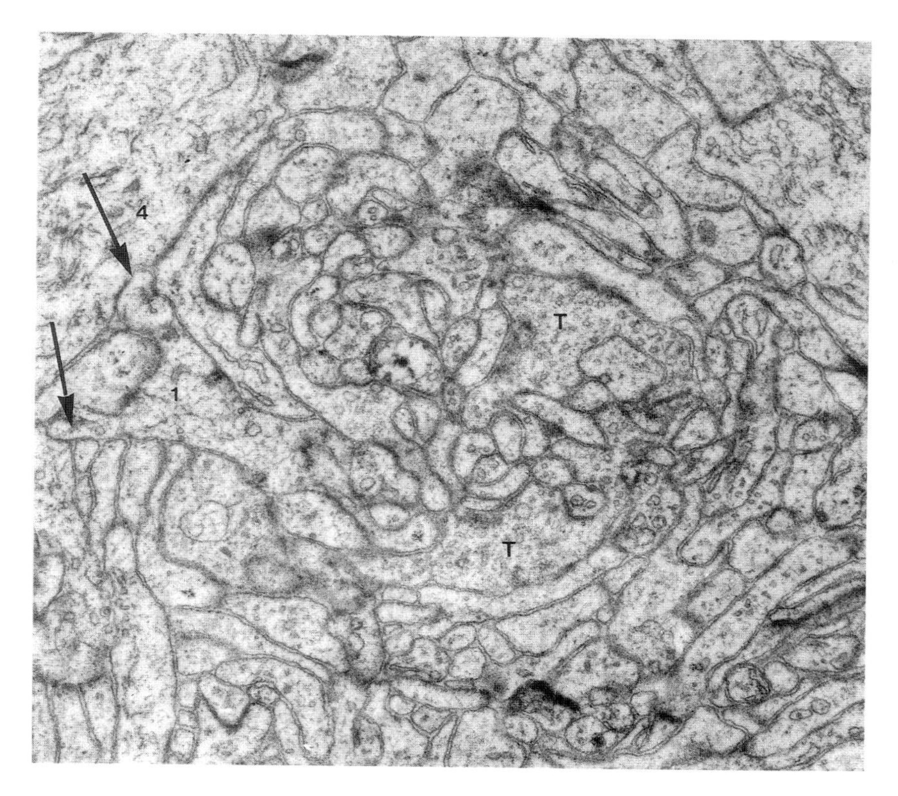

Fig. 1.26. Section 45. The subsynaptic neuropil dominates at this level, with only small areas covered by the terminal (T). A widened part of bipolar cell 1 is located to the left. Two collateral branches (arrows) extend to the left. One contacts horizontal cell 4; the other contacts horizontal cells 34 (small horizontal cell) and 4 (large horizontal cell) further in a vitread direction. 23,000×.

foreground and the montages of electron micrographs of each section are hanging to the left. Figures 1.35 and 1.36 show the model in closer views, and Figs. 1.37 and 1.38 are drawings of the model. This model was built during the first phase of the extended analysis.

The neural endings contacting the terminal are located in the lower middle part of the model where the reconstructed endings and neural processes are densely arranged. The outer plexiform layer is located above the terminal. Thus the vitread direction is upward. This orientation will be retained in the pictures of the models but it will be reversed in the schematic drawings, where the orientation therefore will be the same as that in Figs. 1.15 and 1.16.

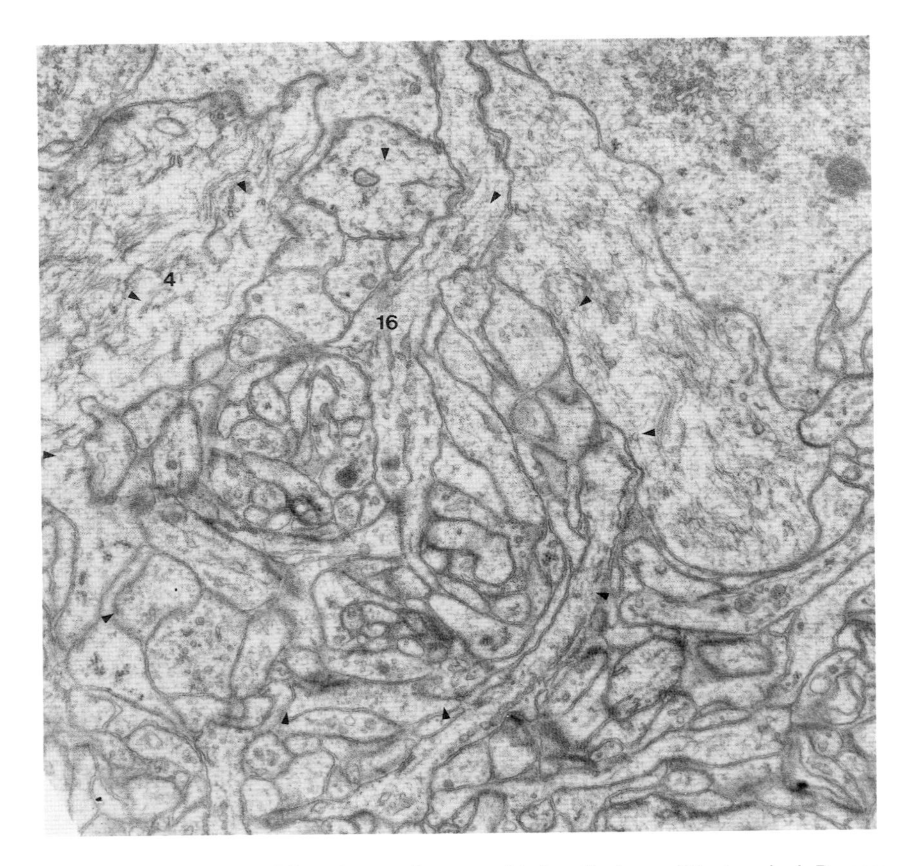

Fig. 1.27. Section 42. The subsynaptic neuropil below the base of the terminal. Process 16 belongs to a small horizontal cell. Its diameter can be compared to that of the process of the large horizontal cell 4. Arrowheads outline an area that corresponds approximately to the main body of terminal 1 projected onto the plane of this section. 23,000×.

The densely arranged end branches in the lower middle part of Fig. 1.38 are all located in deep invaginations of the plasma membrane of the terminal. Figures 1.39 and 1.40 are higher magnifications of the invaginated endings at the terminal. The synaptic ribbon complexes are indicated by the inclusion of the ribbon structure. The ribbons appear white in the photographs, and in Figs. 1.39 and 1.40 they are labeled by letters. The ribbon complexes are located at several levels in the vitreo–sclerad direction. The pictures reveal the depth and the complex shapes required for the invaginations of the photoreceptor membrane to accommodate the horizontal cell and bipolar cell endings. The pictures

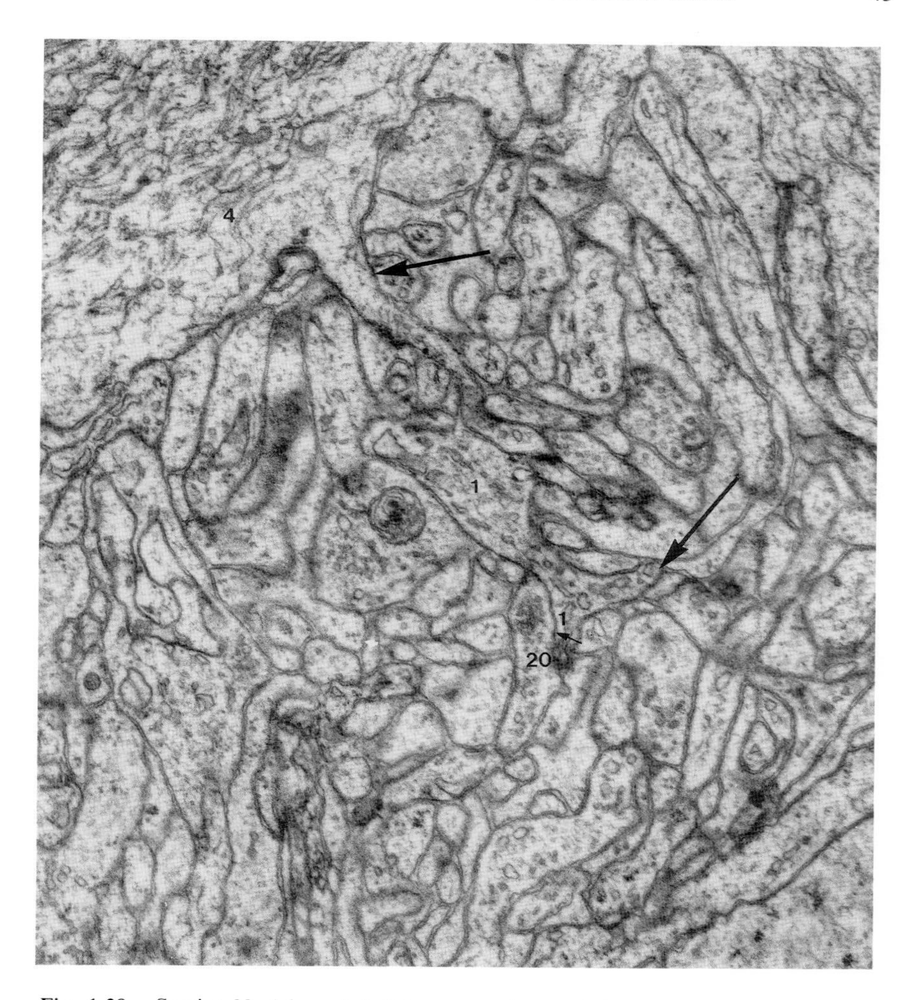

Fig. 1.28. Section 39. A branch (left large arrow) extends from horizontal cell process 4 to the subsynaptic neuropil and contributes endings to six synaptic ribbon complexes. It is characteristic of the large horizontal cells to be connected to the terminals by such short and thin branches that extend directly from the thick trunk of the main process. The thick part of the bipolar cell 1 process divides into one collateral and an end branch (right thick arrow) that can be traced to a rod terminal. This collateral contacts a process from the small horizontal cell 20 (thin arrow). An accumulation of synaptic vesicles in the latter process. 23,000×.

also show that the bulbous horizontal cell endings are completely separated by the ribbon within each complex, and the ribbons therefore reveal the approximate height of the septum that is interposed between

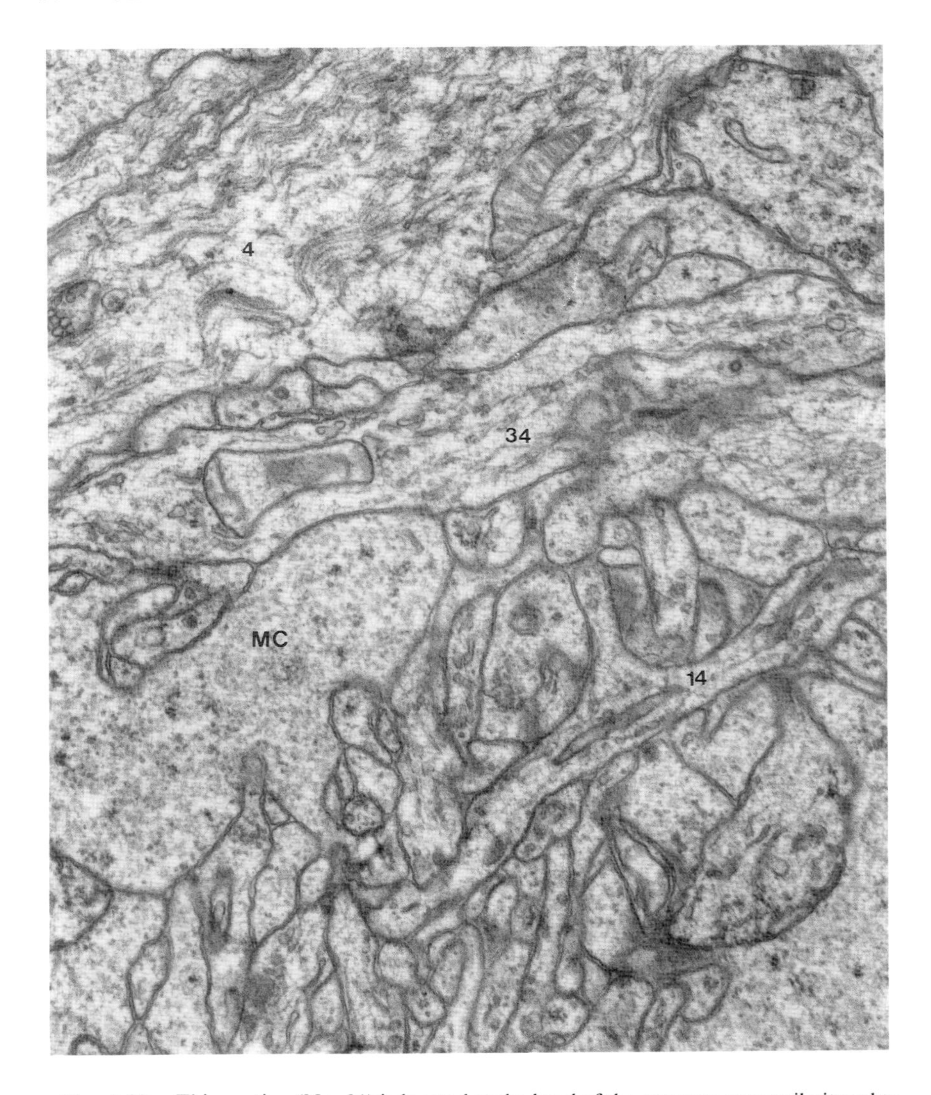

Fig. 1.29. This section (No. 31) is located at the level of the common neuropil vitread to the subsynaptic neuropil. Process 34 is a small horizontal cell process that can be compared to the large horizontal cell 4 process in the upper left corner. A Müller's cell process (MC) occupies a large area. 22,000×.

the widened horizontal cell endings. Characteristically, individual sections do not convey the true depth to which the invaginations reach or the true height of the septum containing the synaptic ribbon. This has

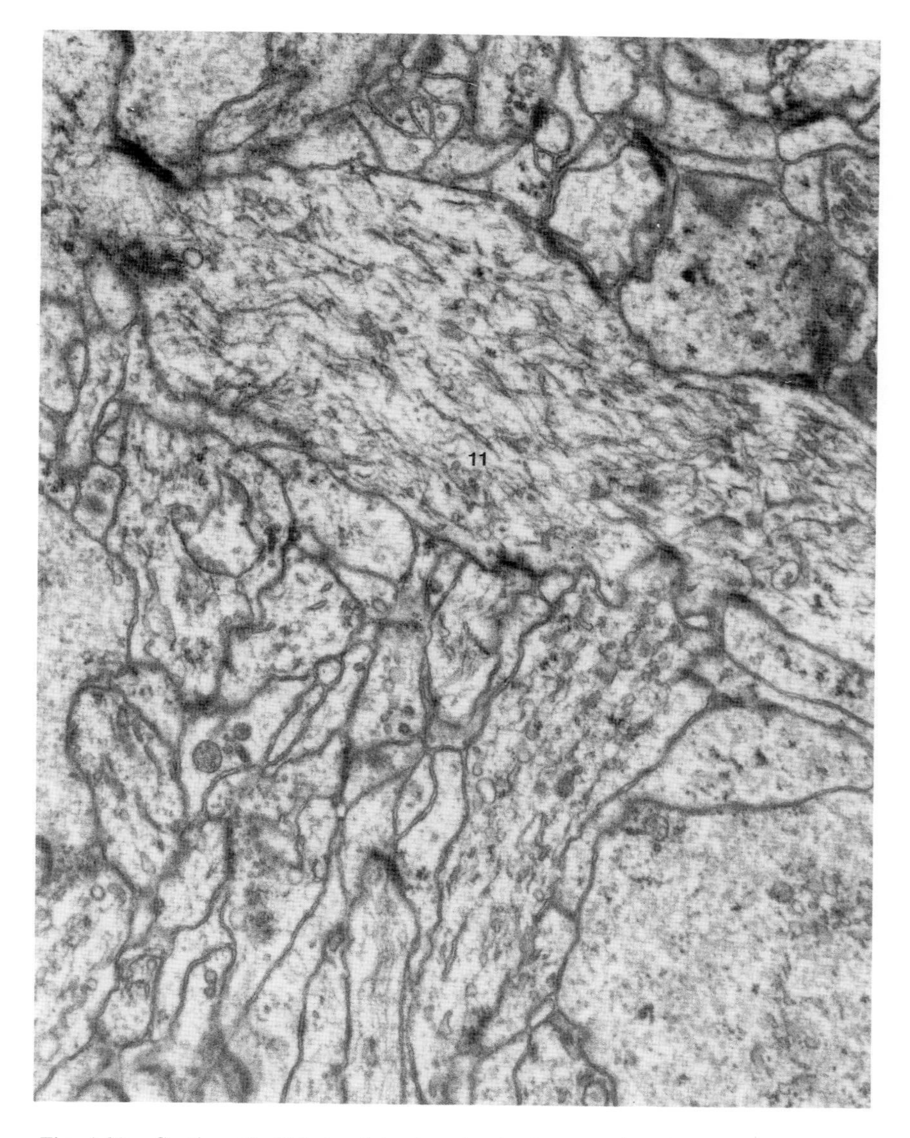

Fig. 1.30. Section 13. This level is close to the inner nuclear layer. The terminal is located above large horizontal cell process 11 and its center corresponds to the location of figure 11. 22,000×.

contributed to the misconception that the bipolar cell endings are located centrally and that the horizontal cell endings are located laterally.

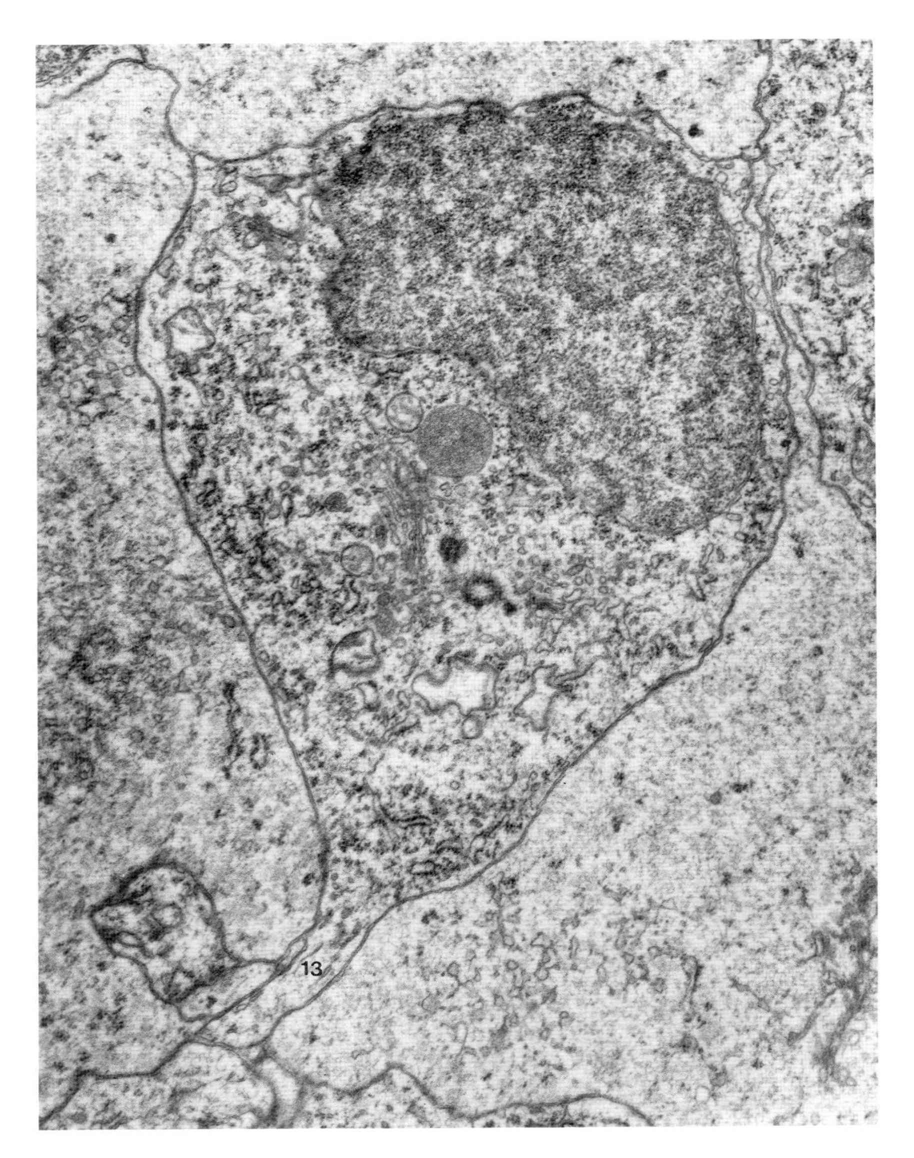

Fig. 1.31. Section through the inner nuclear layer showing the soma of bipolar cell 13 that contacts terminal 1. 19,000×.

The circuitry contained within the volume of neural connections shown in Figs. 1.39 and 1.40 is not included in what is generally referred to as the outer plexiform layer. A major part of the circuitry involving the

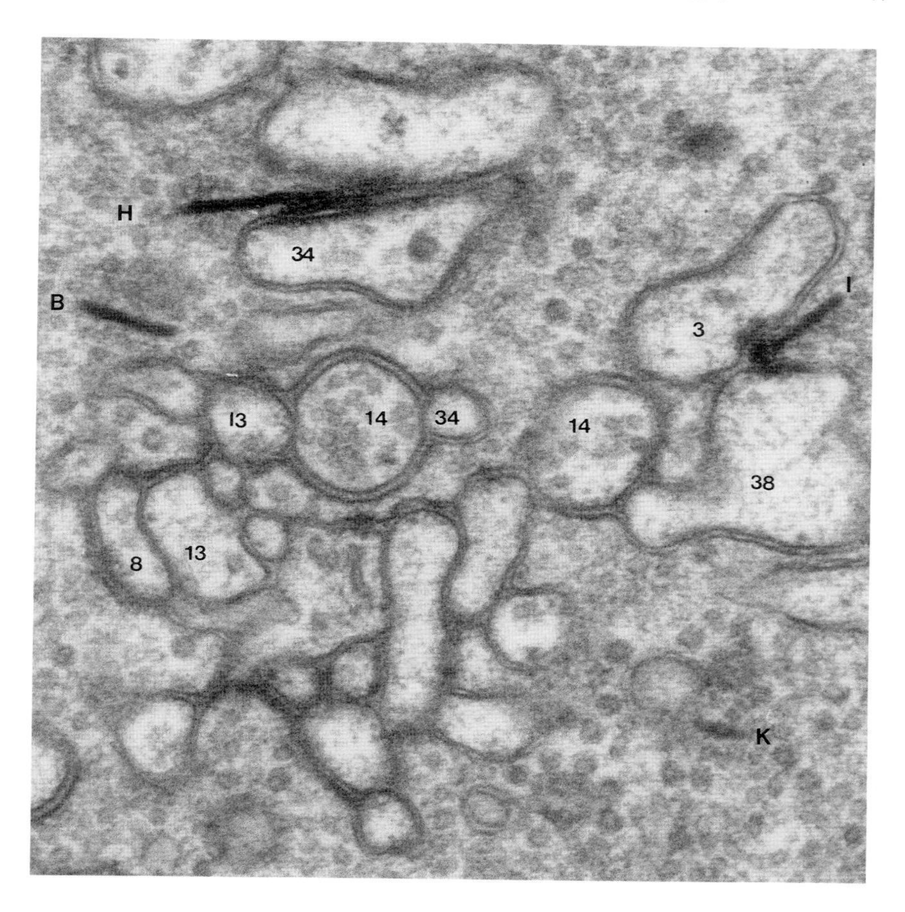

Fig. 1.32. Higher magnification (9.7 times light optical magnification) of an area of Fig. 1.20, demonstrating that an electron optical magnification of 6400× is sufficient to show all the structural information necessary for this kind of analysis. In the middle of the picture are two cross-sectioned endings of bipolar cell 14. At this magnification the differences between the apposed membranes at these endings and those at horizontal cell endings 3 and 34 can be fully appreciated. Indications of spokelike bridges between the two neural membranes can be seen at the left ending of bipolar cell 14. The membranes of bipolar cells 8 and 13 show a close contact. The profile of the right ending of bipolar cell 14 is in lateral contact with horizontal cell 38. The neural membranes at the contacts between the two endings of bipolar cell 14 and horizontal cells 38, 34, and 13 can be compared with those of the adjacent contact between bipolar cell 14 and the cone membrane. The definition of the structure can be greatly improved without increasing the electron optical magnification and still confining section staining to uranyl acetate by reducing the thickness of the section from 500 to a few hundred Ångströms. Electron optical magnification, 6400×; final magnification, 62,000×.

Fig. 1.33. An early stage in the construction of the model of terminal 1.

Fig. 1.34. The completed model of terminal 1 and its connections. The light table used for the tracings is in the foreground, and the montages of electron micrographs of each section are hanging to the left.

Fig. 1.35. A northwest view of the model of terminal 1 and its connections. Bipolar cell 1 is marked by alternating blue and brown plastic spacers. The vitread direction is upward.

Fig. 1.36. The model seen from the south. The connections at the terminal are in the lower center of the picture, and the reconstructed outer plexiform layer extends above these connections. The white pieces of plastic indicate the synaptic ribbons. It is at the synaptic ribbons that the neural endings make connections with the terminal and the endings are received in deep invaginations. Because the outer plexiform layer is located above the terminal, the orientation is opposite to that shown in the cross sections of the retina, for instance, in Fig. 1.15.

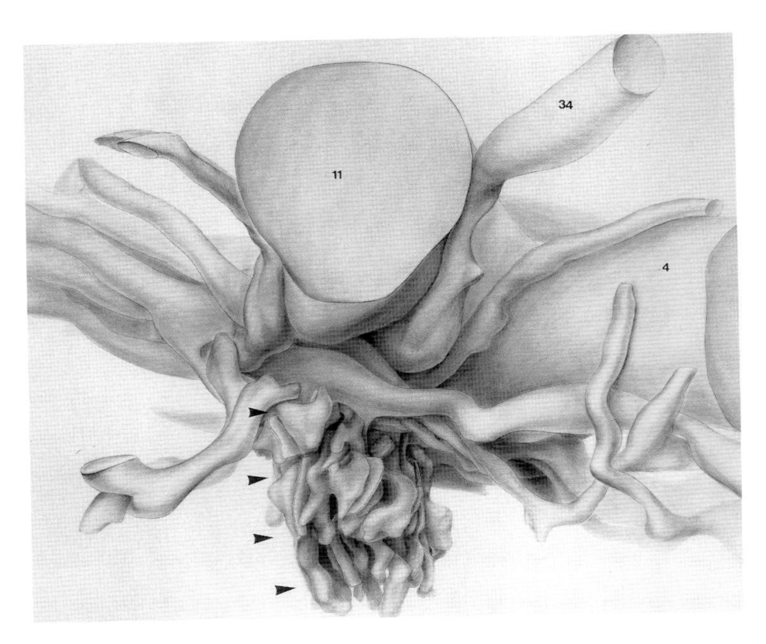

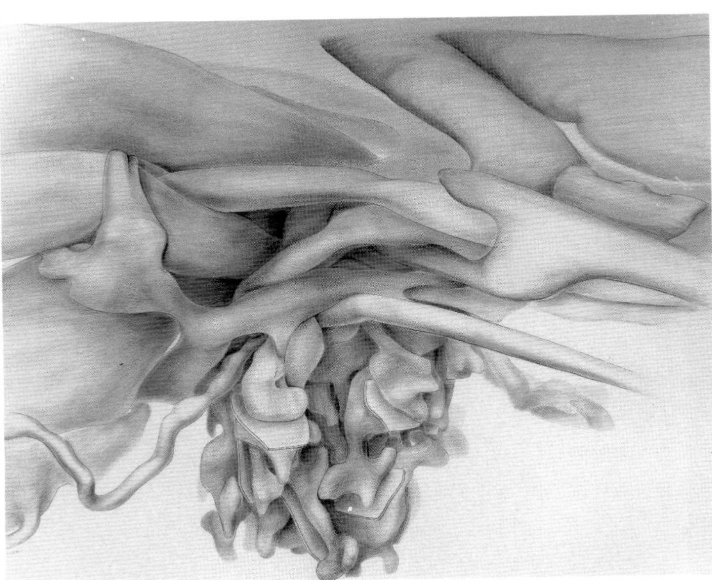

Fig. 1.37 and 1.38. Drawings of the model seen from the north (Fig. 1.37) and from the south (Fig. 1.38). Processes 11 and 4 in Fig. 1.37 are contributed by two large horizontal cells, while process 34 is contributed by a small horizontal cell. The difference in the thickness can be appreciated as well as the variation of the thickness of the small horizontal cell process along its course. Arrowheads in Fig. 1.37 indicate the location of the invaginated endings. Drawings by Mrs. Hermine Kavanau.

Fig. 1.39. A closer view of the neural endings at the terminal viewed from the east. The flat sheets indicate the locations of the synaptic ribbons. The synaptic ribbon complexes, indicated by letters, are present at several levels of the terminal; compare, for instance, the locations of complexes A and P. The voluminous endings of the horizontal cells dominate among the endings at the terminal. Ending 6 is a horizontal cell ending associated with two adjacent synaptic ribbon complexes, A and F.

photoreceptors, the bipolar cells, and the horizontal cells is, in fact, located within the invaginations at the photoreceptor membrane and outside the outer plexiform layer.

Anatomical models are helpful for establishing precise three-dimensional topographic relationships that may sometimes be difficult to

Fig. 1.40. Another close view of the connections at the terminal as seen from the south. Horizontal cell 11 is connected to two synaptic ribbon complexes, F and O, in series. The two widened parts of the ending at the two synaptic ribbons are connected by a thin segment. Bipolar cell 1 dendrite passes through the subsynaptic neuropil, and one collateral is seen to branch off, connecting large horizontal cell 4 and small horizontal cell 34 (arrows).

determine by viewing electron micrographs of consecutive sections. Anatomical models were therefore used during the second phase of the project (Sjöstrand, 1978) in situations where the topographic relationships were particularly complex (Fig. 1.41). Reconstructions were then made at 70,000× magnification, which is the lower limit of magnification useful in

Fig. 1.41. Smaller scale anatomical model built as a supplement to the linear model to analyze particularly difficult topographic relationships during the reconstruction of terminal 2.

Fig. 1.42. Pieces of plastic cast used for building linear models. From Sjöstrand (1974).

Fig. 1.43. Lower portion of the linear model of terminal 2 illustrating the coordinate system according to which the model was built. Slightly off center are connections to terminal 2 (arrow). At the lower right corner are neural connections to two rod terminals. The grill work represents horizontal cells and shows the diameter in the vitreo–sclerad direction of these processes. From Sjöstrand (1978).

study of the structure of photoreceptor terminals. This size of the reconstruction is not convenient for analysis because the processes in this part of the retina are so minute.

Otherwise, the reconstruction during the second phase of the project involved a linear model built from standardized pieces of plastic of different colors (Fig. 1.42) that were glued together; in this case the prints of the electron micrographs were placed on a coordinate system drawn on the top of the light table. The position of each process relative to this coordinate system was determined, and the model was built on a corresponding coordinate system. This is illustrated in Fig. 1.43, which shows part of the model of the second cone terminal.

So that the linear model could be viewed, it was built in sections. Each model section was reconstructed from a certain number of microtome sections, the number depending upon how densely the processes were

Fig. 1.44. The complete linear model of connections to terminal 2 and to surrounding rod terminals. The upper part of the model extends into the inner nuclear layer, with the soma of neurons contributing endings to terminal 2. From Sjöstrand (1978).

arranged. When the arrangement was particularly dense, each model section represented only four or five microtome sections so that the intertwined arrangement of the processes could be viewed without being obscured by too much superpositioning of processes.

Fig. 1.45. The linear model after some model sections have been removed. Each section is built on a separate sheet of plastic onto which the coordinate system has been drawn. The model can therefore be built with reference to the same coordinate system through its entire height. Plastic spacers at the corners of the plastic sheets allow the model sections to be stacked on top of each other. A part of the model that was built before the final technique was developed is in the background.

Fig. 1.46. Close view of the connections at terminal 2, with the subsynaptic neuropil in the center and some synaptic ribbon complexes to the right.

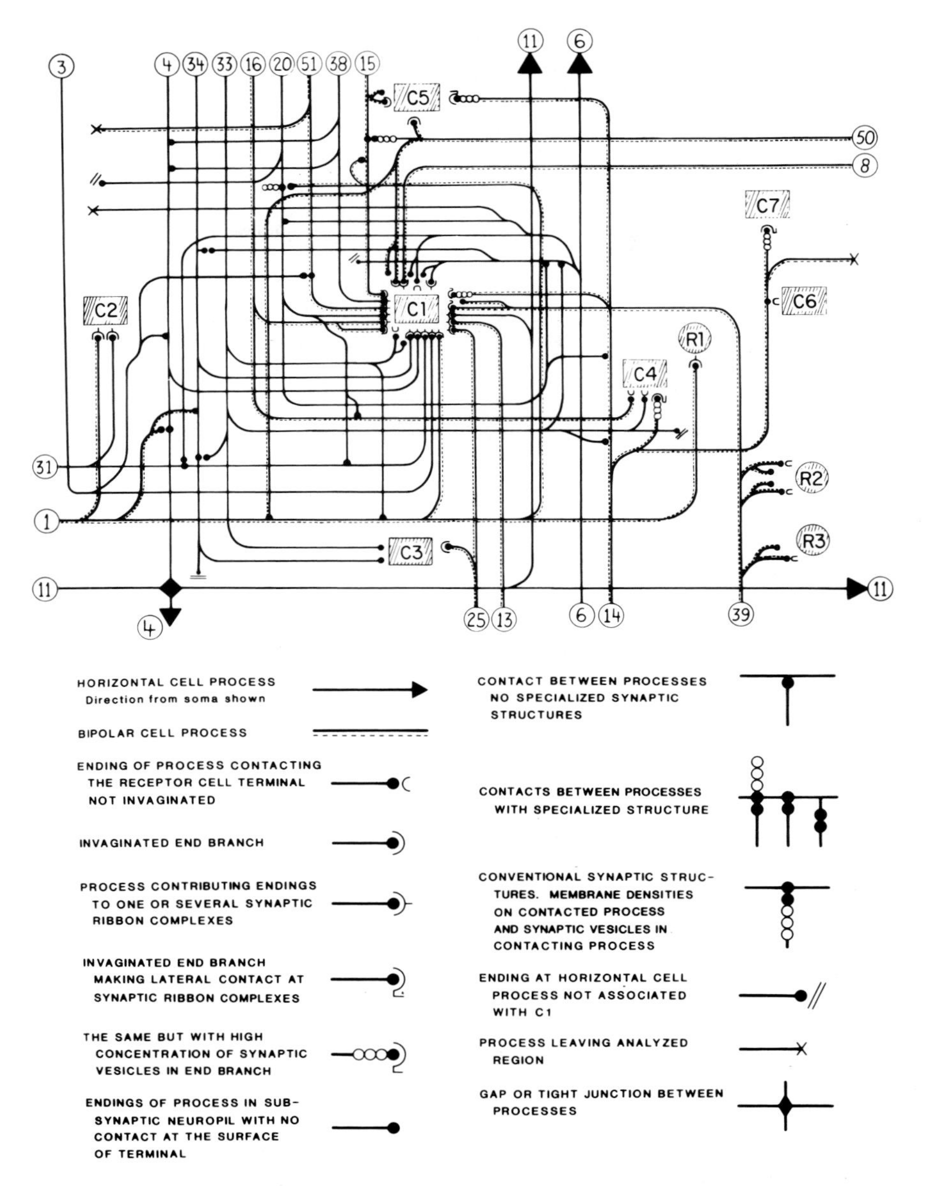

Fig. 1.47. The extent of the three-dimensional reconstruction and of the tracings done during the first phase of analysis of neural connections in the reconstruction of terminal 1. The approach directions of the neural processes are shown as well as the relative positions of the terminals to which the neurons connected to terminal 1 also connected. The bipolar cell dendrites were connected predominantly to cone terminals (CT), but three out of nine bipolar cells (1, 25, and 39) were also connected to rods within the reconstructed region. The type of connection—not invaginated connections at the surface of the terminal, invaginated connections, synaptic ribbon connections, lateral connections to synaptic ribbon complexes—is also shown. From Sjöstrand (1974).

The finished model is shown in Fig. 1.44, and Fig. 1.45 illustrates how various parts of the model can be analyzed by removing model sections to expose neural patterns in the interior of the model. Figure 1.46 is a close-up view of the neural processes in the most sclerad part of the terminal.

The building of such a linear model is not time-consuming and satisfies most requirements; in any case no model can replace the electron micrographs. It is only when topographic relationships are particularly complex that an anatomical model is required. An anatomical model covering a limited region can then be built. The second terminal was reconstructed and partially analyzed in 18 months, including time spent on developing the technique for building a linear model.

Figure 1.47 shows the tracings done during the reconstruction of terminal 1 only. It reveals the relative positions of the photoreceptor terminals and the approach directions of most of the neurons that were traced in the reconstruction of terminal 1.

Chapter 2: The Neural Connections at Rod and Cone Terminals

1. The Synaptic Ribbon Complexes at Cone Terminals

The sixteen synaptic ribbon complexes at terminal 1 offered a good opportunity to reveal certain basic features and variations of the structure of these complexes. The fact that certain structural features are always present points to a particular functional significance of these features. This significance will be discussed later, when the function of the synaptic ribbon complexes is deduced, in Chapter 3.

Some synaptic ribbon complexes are shown in Figs. 1.39 and 1.40, and the arrangement of the neural endings is more clearly revealed in Figs. 2.1 and 2.2. In most cases only a single bipolar cell ending contacts each complex, forming a triad structure with two horizontal cell endings. The pictures clearly illustrate the difference between the dimensions of the bipolar cell endings and those of the horizontal cells. In contrast to pictures of individual sections, the model revealed the considerable depth of the invagination and of the height of the cytoplasmic septum containing the synaptic ribbon that separates the endings of the two horizontal cells. The latter endings are obviously located predominantly sclerad to the bipolar cell ending.

At the most vitread part of the synaptic ribbon complexes, the horizontal cell end branches are thin, with roughly the same diameter as that of the bipolar cell ending. They then widen abruptly to form voluminous endings located in the sclerad part of the complex. This basic organization of the synaptic ribbon complexes is illustrated schematically in Figure 2.3.

The same horizontal cell end branch sometimes contacts two synaptic ribbon complexes in series, with one complex located more vitread than the other. In such cases, two widened parts of the end branch are

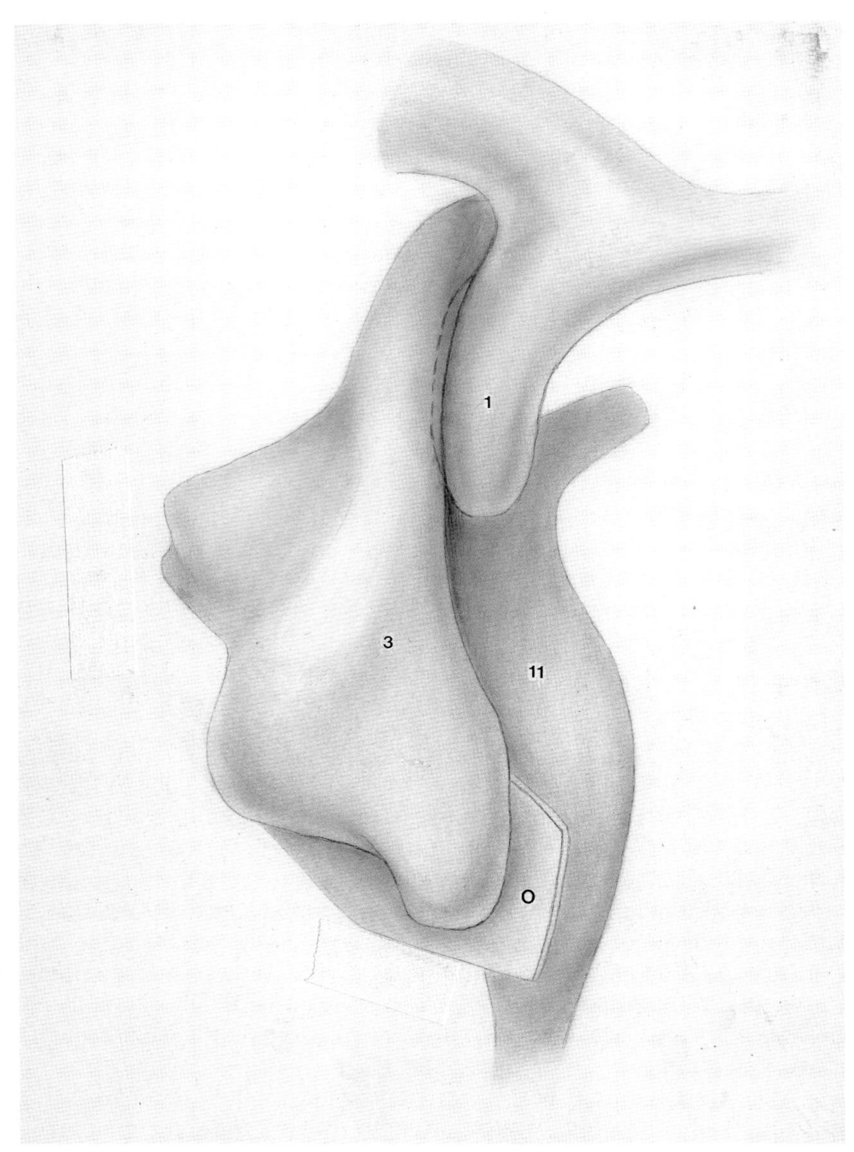

Fig. 2.1. Reconstructed synaptic ribbon complex (O) at terminal 1, also shown in Fig. 1.40. Vitread direction upward in Figs. 2.1 and 2.2, opposite to that in the schematic drawings. Two widened parts of horizontal cell endings, 3 and 11, are separated by a synaptic ribbon, O, whereas one short, thin end branch from bipolar cell 1 contacts the narrow parts of the two horizontal cell endings. This contact is located vitread to the widened parts. Horizontal cell ending (11) extends further in a sclerad direction (downward) and then ends, with a second widened part contributing one of the horizontal cell endings to a second synaptic ribbon complex. From Sjöstrand (1974).

associated with the two synaptic ribbon complexes, and these widened parts are connected by a thin part. Figure 1.40 shows one such case involving a horizontal cell marked 11, and Figure 2.4A illustrates these relationships schematically. Horizontal cell 11 is paired with different horizontal cells at the two ribbon complexes, and it is also contacted by different bipolar cells at the two complexes. The second complex is structurally basically identical to the first complex because at both complexes the narrow parts of two horizontal cell endings are in a triad type of contact with a bipolar cell ending in the vitread part of the complex, while the widened parts extend in a sclerad direction beyond this triad contact.

Figure 2.2 illustrates the participation of one horizontal cell ending at two synaptic ribbon complexes arranged in parallel. The thin part of the horizontal cell ending in this case passes by the endings of the two bipolar cells that contact the two synaptic ribbon complexes before it widens to form one bulbous ending that is shared by the two synaptic ribbon complexes. The thin part of the horizontal cell ending thus contributes to a triad type of contact at both complexes. There are different bipolar cells and different horizontal cells that join this horizontal cell ending at the two complexes. This type of connection is illustrated schematically in Fig. 2.4B.

The analysis made it possible to distinguish between two structurally different parts of the complex. One sclerad part ("proximal," according to Sjöstrand's (1958) nomenclature) is characterized by the widened bulbous parts of the two horizontal cell endings being exclusively and individually in contact with the photoreceptor. Any mutual contact between the two endings is prevented by the interposed cytoplasmic septum containing the synaptic ribbon.

In the vitread ("distal") part of the complex the situation is entirely different. Here the narrow parts of the two horizontal cell endings are in mutual contact, and both endings are also in contact with the bipolar cell ending which does not extend sclerad any further. It is in this vitread part that the triad arrangement applies. When two bipolar cell endings are involved, the endings are lined up in series along the narrow parts of the two horizontal cell endings (Fig. 2.5).

The endings in the vitread part measure only 0.1–0.15 μm in diameter, and being closely associated they occupy a small volume in comparison to the sclerad part.

While the sclerad part is characterized by horizontal cell–photoreceptor connections, the vitread part is characterized by the quadruple or quintuple contacts between two horizontal cells, one or two bipolar cells, and the photoreceptor.

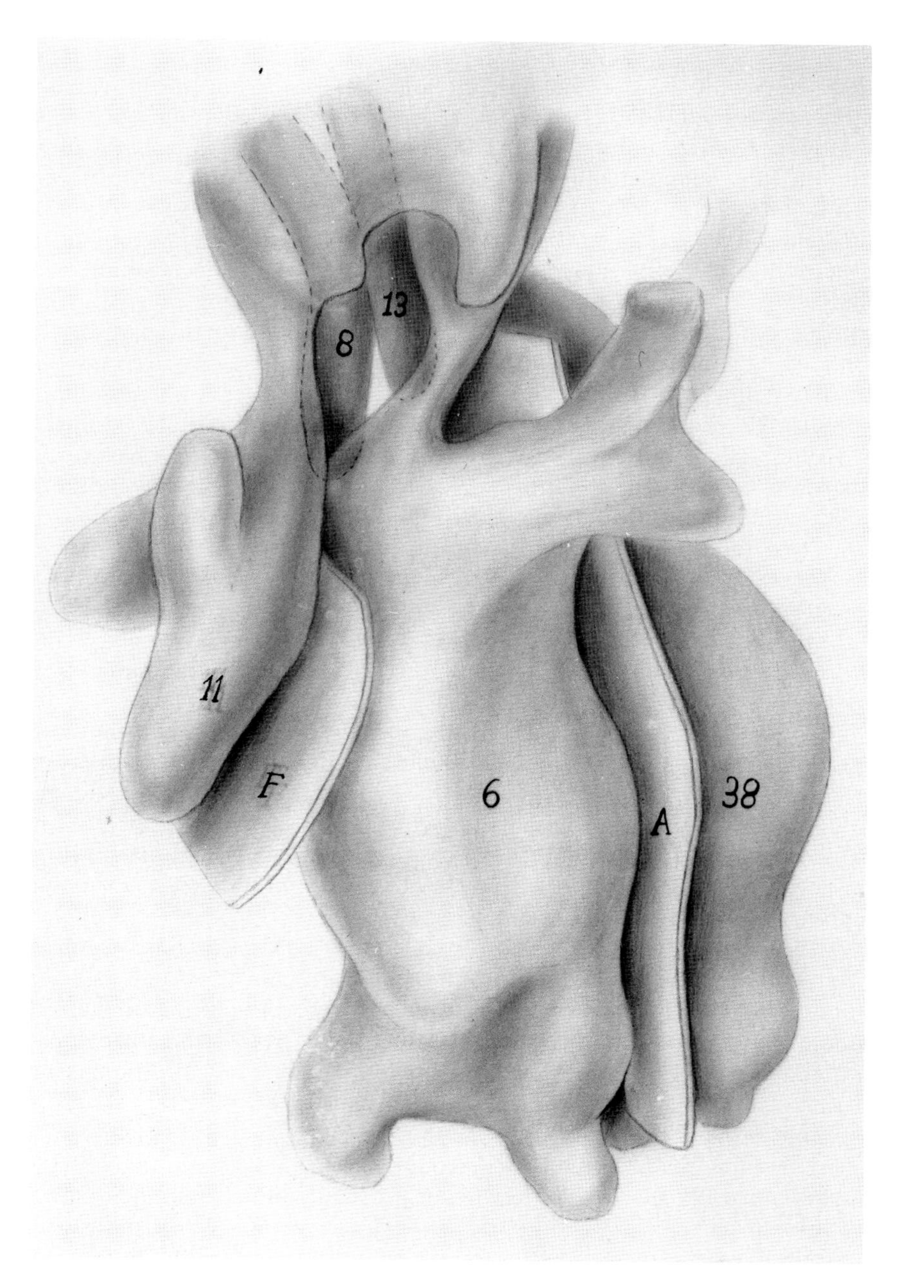

Fig. 2.2. Two synaptic ribbon complexes (A and F), also shown in Fig. 1.39, at which one horizontal cell ending (6) is shared by both complexes. Two bipolar cell endings (8 and 13) contact the horizontal cell endings at its narrow part. This is possible because the narrow part of horizontal cell 6 extends obliquely from the vitread part of complex A to the vitread

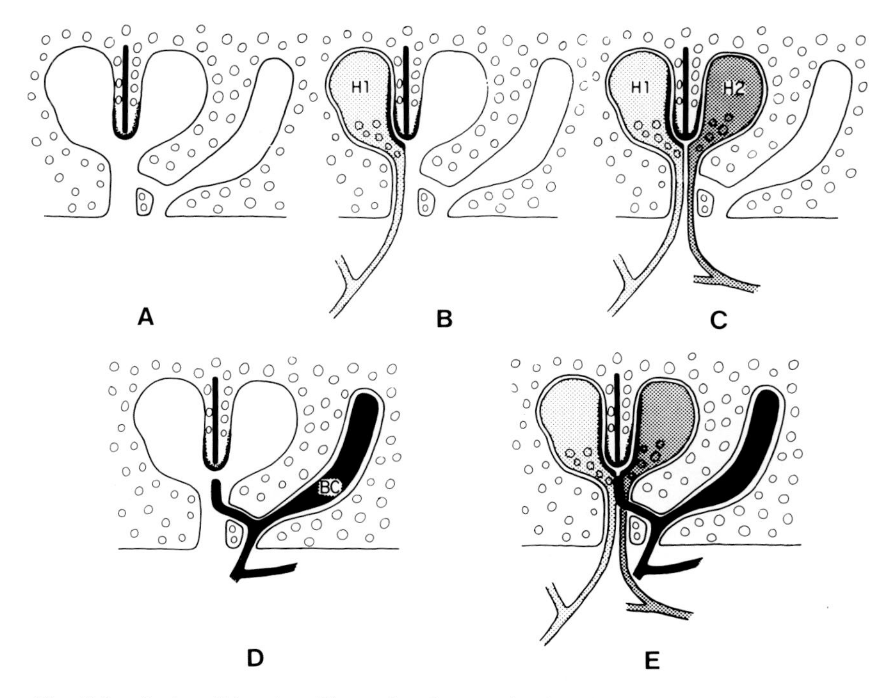

Fig. 2.3. Series of drawings illustrating the organization of synaptic ribbon complexes in cone terminals. (A) The invagination of the basal surface of the plasma membrane of the terminal is shown. The invagination consists of a narrow entrance and branches into two large chambers separated by a septum in which the synaptic ribbon is located. (B) One horizontal cell ending fills the left chamber. (C) A second horizontal cell ending fills the right chamber. (D) The ending of a bipolar cell located in an invagination extending laterally relative to the synaptic ribbon complex. One thin, side branch contacts the horizontal cells at the synaptic ribbon complex in its vitread part. (E) All endings are in position. Synaptic vesicles are present both in the terminal and in the horizontal cell endings. There is one zone extending 0.2 μm from the edge of the septum that is free of synaptic vesicles. Cytoplasmic densities are present at the septum at both the photoreceptor and the horizontal cell membranes. Vitread direction downward.

---

part of complex F. It is paired with the narrow part of horizontal cell 38 at complex A to form a triad connection with bipolar cell ending 13. Because of its oblique course, the ending pairs with horizontal cell 11 in complex F, forming a triad connection with the bipolar cell 8 ending at this complex. This arrangement, as well as that shown in Fig. 2.4, reveals that the triad contacts involve the narrow part of the horizontal cell endings before these processes widen to form the large bulbous sclerad endings. The triads are located in the vitread part of the complex, whereas the widened endings of the horizontal cells extend into the sclerad part where a septum in which the synaptic ribbon is located completely separates the two horizontal cell endings. From Sjöstrand (1974).

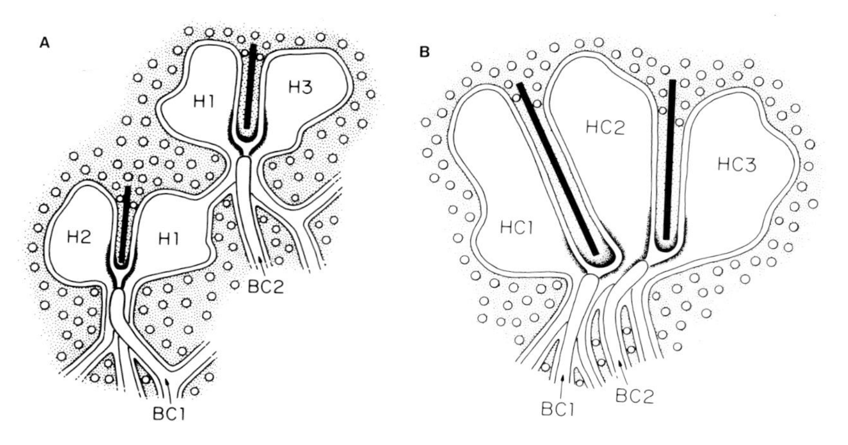

Fig. 2.4. Multiple connections to synaptic ribbon complexes by a single horizontal cell ending. (A) One horizontal cell ending makes connections at two synaptic ribbon complexes arranged in series. (B) A corresponding double connection to synaptic ribbon complexes arranged in parallel.

There is one basic difference with respect to the connections of the bipolar cells that contributed end branches to the synaptic ribbon complexes. The bipolar cell end branches either are a terminal ending of a bipolar cell dendrite or are side branches extending from terminal endings that are received in special invaginations of the photoreceptor membrane. In the latter cases the terminal endings thus establish connections with the photoreceptor terminal separate from the endings at the synaptic ribbon complexes. This difference is illustrated schematically in Fig. 2.6.

The special terminal branches to the photoreceptors outside synaptic ribbon complexes are characteristic of certain bipolar cell connections at cone terminals.

There are considerable differences in the structure of the neural membranes between the vitread and sclerad parts. The neural membranes of the two horizontal cell endings are particularly thin in the vitread part, and the membrane of the bipolar cell ending is also particularly thin. The space separating these membranes is narrow.

Cytoplasmic densities are associated with the neural membrane of both horizontal cell endings and the photoreceptor membrane. In the horizontal cell endings these densities are particularly conspicuous at the part of the horizontal cell membrane that faces the septum containing the synaptic ribbon (see for instance Fig. 2.7). Such areas, however, may also be present at other parts of these endings facing the terminal.

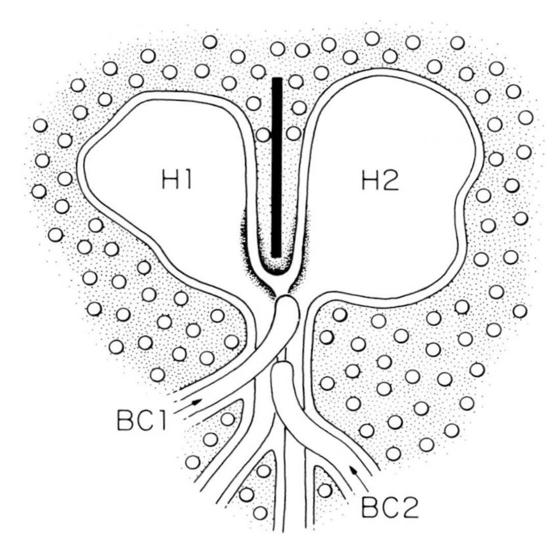

Fig. 2.5. Synaptic ribbon complex with two bipolar cell endings connected to two horizontal cells. The bipolar cell endings are lined up along the thin part of the horizontal cell ending. This arrangement applies only to cone terminals.

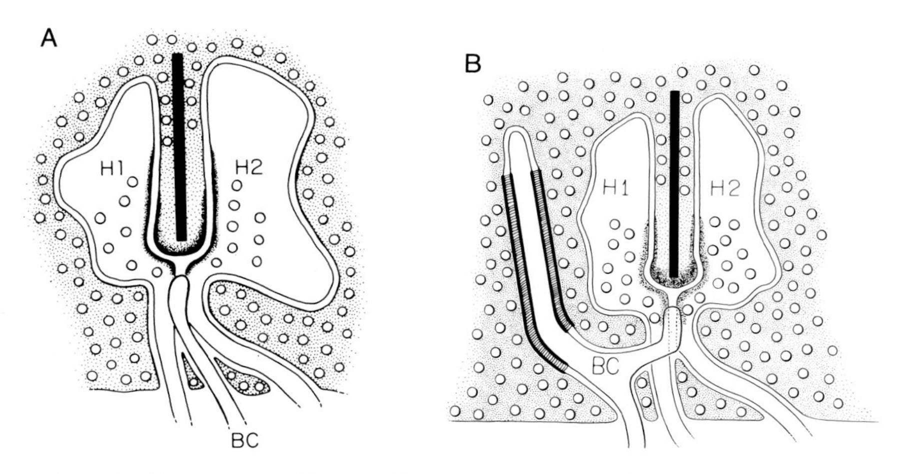

Fig. 2.6. The two ways bipolar cells connect to synaptic ribbon complexes: (A) by a terminal ending and (B) by a side branch from the terminal ending that is received in a separate invagination of the cone membrane.

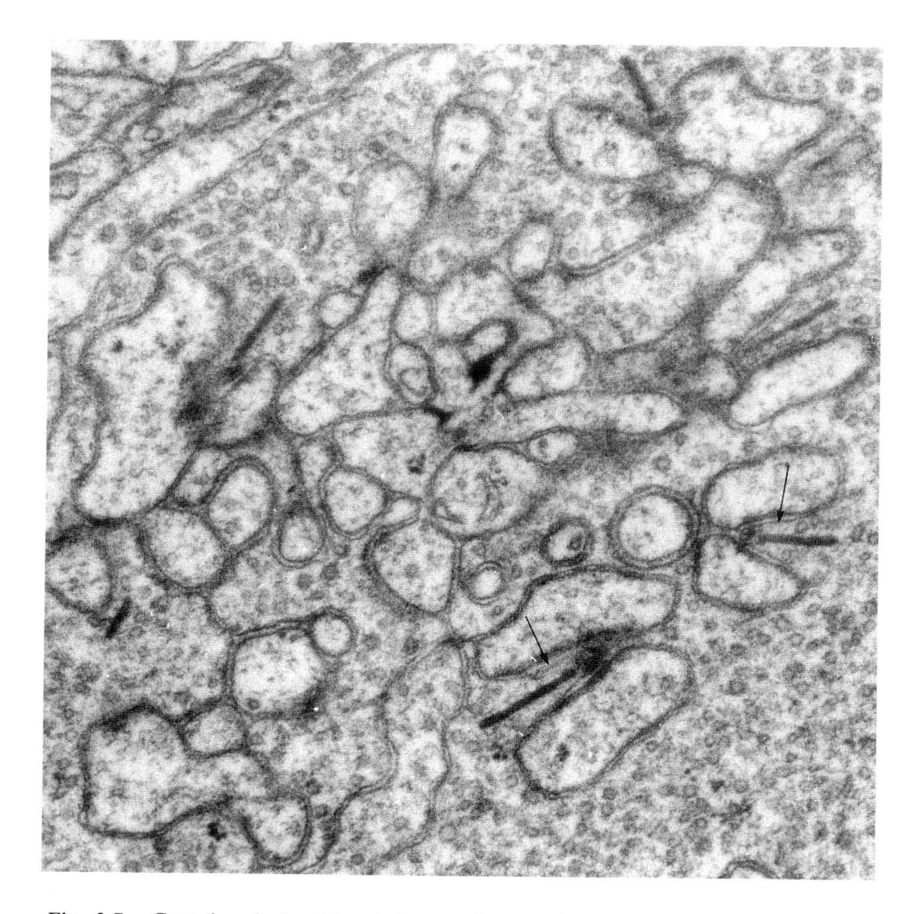

Fig. 2.7. Cytoplasmic densities at the neural connections of synaptic ribbon complexes. Such areas are present both at the horizontal cell membrane and at the photoreceptor membrane. Note the absence of synaptic vesicles in a zone at the edge of the septum containing the synaptic ribbon (arrows).

Cytoplasmic densities are present at the photoreceptor membrane near the edge of the septum (Sjöstrand, 1958) but may also include other restricted areas at the sclerad part of the invagination. The edge of the septum faces the part of the extracellular space where horizontal cell and bipolar cell endings face each other. That the neural membranes of the horizontal cells and of the terminal have a special structure in this region has been confirmed by an analysis of freeze-fractured retinas (Raviola and Gilula, 1973).

The horizontal cell end branches contain synaptic vesicles that are frequently located predominantly at and close to the vitread part of the end processes. On the photoreceptor side there are no synaptic vesicles in a zone within the septum that separates the two widened parts of the horizontal cell endings. This zone extends 0.2 μm vitread from the edge of the septum, as illustrated in Figs. 2.6 and 2.7.

These observed structural differences between the vitread and sclerad parts of the complex indicate that the two parts of the synaptic ribbon complex fulfill different functions. While the sclerad part structurally favors interaction between photoreceptor and horizontal cells, the vitread part allows interaction between the photoreceptor, the horizontal cells, and one or two bipolar cells.

2. Are the Neural Connections at the Synaptic Ribbon Complexes Functional Neural Connections?

While nobody seems to have questioned that connections between neurons as revealed by light microscopic analysis of Golgi-stained preparations constitute functional connections, the functional significance of the connections revealed by the electron microscope has been questioned, because neuroanatomists accustomed to a coarse and oversimplified light microscopic version of the structure of the nervous system presumably are overwhelmed by the number of neural processes and connections and the minuteness of the processes.

The problem of evaluating neural connections will be approached in Chapter 3, Section 28. Here we only point out the features of the synaptic ribbon complexes that make it clear that we are dealing with functional neural connections.

That the connections are functional is obvious for several reasons. The horizontal cell-photoreceptor connections at the synaptic ribbon complexes are the dominant and frequently only type of connection between these two types of neurons. With respect to the bipolar cell endings at synaptic ribbon complexes it seems obvious that they are functional connections because these endings are either furnished by special side branches or constitute the only type of bipolar cell ending at the terminal. Figure 2.8 shows the thin terminal branches of bipolar cell 1 at terminal 1 contacting five synaptic ribbon complexes. One of the branches makes an en passant connection at synaptic ribbon complex K. Three larger terminal branches are received in separate invaginations of the photoreceptor membrane.

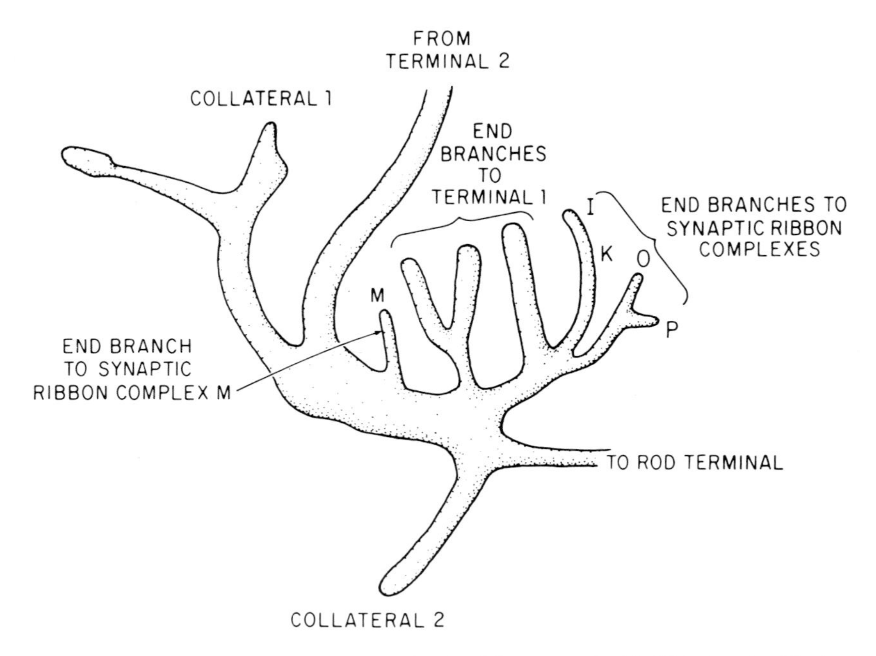

Fig. 2.8. Branchings of the bipolar cell 1 dendrite to terminal 1. This drawing illustrates the various types of endings contributed by bipolar cell 1 at terminal 1. There are four endings to five synaptic ribbon complexes, one ending participating in the triad at two complexes (I and K). Three endings are received in deep invaginations and constitute separate connections to the photoreceptor. Two collaterals contact horizontal cells in the common neuropil.

For many bipolar cells the connections made at synaptic ribbon complexes are the only contacts between the bipolar and horizontal cells. Functionally, such contacts must exist for most bipolar cells, and the synaptic ribbon complex is then frequently the only site where functional contacts between bipolar and horizontal cells are possible.

The presence of cytoplasmic densities at the connections between the horizontal cells and the photoreceptor at the vitread part of the synaptic ribbon complexes satisfies one arbitrarily accepted criterion for a synaptic connection. The fact that the horizontal cell endings contain synaptic vesicles and that such vesicles are absent in the photoreceptor terminal within a zone in the septum containing the synaptic ribbon makes it reasonable to assume that the horizontal cells in this region are presynaptic components and that this localized area of the photoreceptor membrane is postsynaptic.

Another feature points to a functional connection between the horizontal cell endings and the bipolar cell endings. The neural membranes are separated by a narrow space and the neural membranes are particularly thin at the triad connections.

3. A First Example of Systematic Patterns of Neural Connections

The large number of synaptic ribbon complexes that were analyzed at the two cone terminals and the large number of neurons contributing endings to these complexes made it possible to determine whether these connections were arranged systematically or randomly. At terminal 1 there were sixteen synaptic ribbon complexes, and altogether almost thirty complexes in cone terminals were analyzed. The large number of horizontal cells and bipolar cells that contacted terminal 1 contributed a total of 49 endings at the 16 ribbon complexes (Fig. 2.9). Seven horizontal cells contributed 24 processes participating at the 16 synaptic ribbon complexes. Some of these end branches contacted two synaptic ribbon complexes in series, and in some cases one horizontal cell ending was shared by two synaptic ribbon complexes.

Of the six bipolar cells that contacted horizontal cells at the synaptic ribbon complexes, four contributed endings to more than one complex. There were 32 horizontal cell connections made by seven different horizontal cells and 18 bipolar cell endings contributed by six bipolar cells at the synaptic ribbon complexes. As a consequence, in a considerable number of cases one bipolar cell made multiple connections to horizontal cells. It could then be established whether these connections involved a random mixing of horizontal and bipolar cells at the complexes or whether the connections were arranged according to some systematic pattern.

The analysis first revealed that the two large endings at the sclerad part of the synaptic ribbon complexes were always contributed by horizontal cells and that the third ending of the triad was always contributed by a bipolar cell. No exception to this rule was found in any of the cone and rod terminals analyzed. This was the first indication of order in the pattern of connections.

It was then established that the two horizontal cell endings at each synaptic ribbon complex were contributed by different horizontal cells, as originally observed by Sjöstrand (1969). This was the second indication of a systematic, nonrandom arrangement of the connections. The deviation from this rule shown by Kolb (1970) in a Golgi-impregnated serially sectioned terminal can be explained by the metal breaking through the

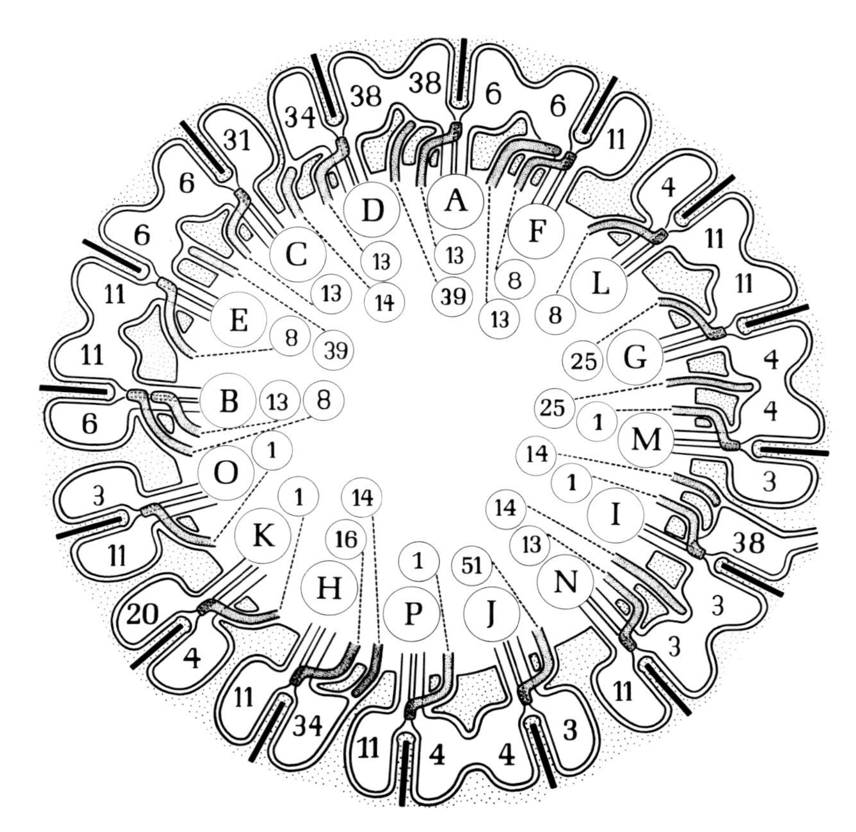

Fig. 2.9. The combinations of bipolar cell–horizontal cell connections at the 16 synaptic ribbon complexes at terminal 1. From Sjöstrand (1974).

thin neural membranes of the two horizontal cell endings, which are in mutual contact at the synaptic ribbon complexes. Such break-through of the metal is one source of error associated with the Golgi technique.

By tracing the horizontal cell end branches to the main process from which they originated and by tracing the main process the direction from which this process approached the terminal was determined. It turned out that the neural processes approached the terminal from four roughly perpendicular directions, which will be referred to as north, south, east and west (Fig. 2.10). The angle between these directions varied between 60 and 90°. The main branches of horizontal cells approached either from north, south, or west. Each horizontal cell process contacting the terminal could consequently be characterized by its approach direction.

A. HORIZONTAL CELLS

4 20 34 33 3

38: APPROACH DIRECTION NOT KNOWN

11
31

TERMINAL 1

6

B. BIPOLAR CELLS

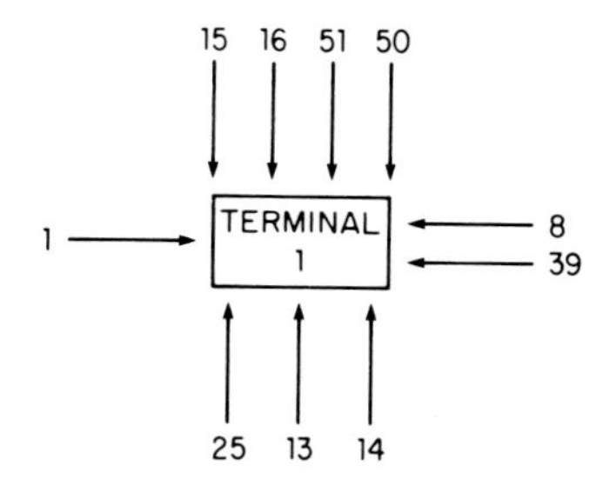

C. LARGE AND SMALL HORIZONTAL CELLS

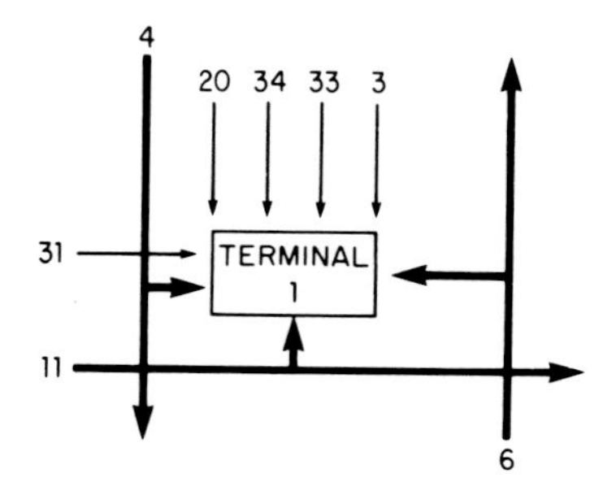

Fig. 2.10. The directions from which neurons contacting terminal 1 approach the terminal.

These approach directions revealed a definite asymmetry in the connections of the horizontal cells to the photoreceptor. No main branch of a horizontal cell approached terminal 1 from east, and only one main branch of a horizontal cell approached from south. Five out of eight horizontal cells approached from the north.

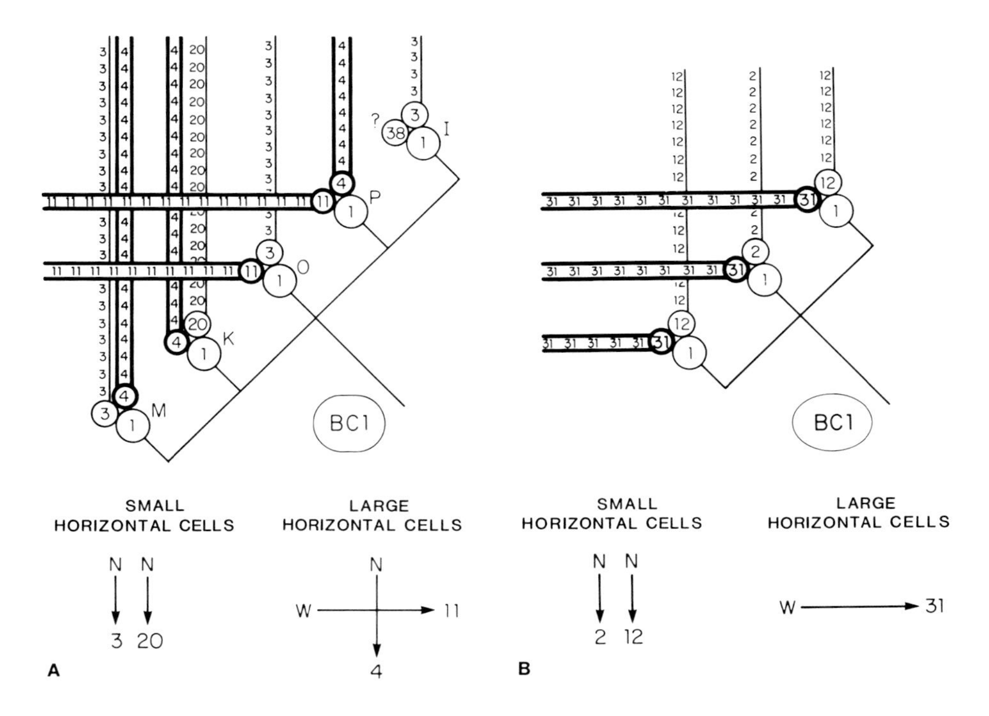

Fig. 2.11. Approach directions of horizontal cell processes connected to bipolar cell 1 at terminal 1 (A) and at terminal 2 (B). The thickest lines indicate large horizontal cells.

The two horizontal cells that contributed endings to the same synaptic ribbon complex approached the terminal from two roughly perpendicular directions in twelve of the fifteen synaptic ribbon complexes in which the approach directions of both horizontal cells could be established. In three complexes, the two horizontal cells approached from the same direction. These combinations of approach directions of horizontal cell processes indicate another systematic arrangement of the neural connections at the synaptic ribbon complexes.

Let us turn now to the connections of the bipolar cells. Characteristically, the bipolar cells that contacted horizontal cells at more than one synaptic ribbon complex contacted horizontal cells that approached the terminal from predominantly the same directions (Figures 2.11–2.13). As an example, bipolar cell 1 contacted four different horizontal cells with known approach directions at four synaptic ribbon complexes at terminal 1. All these horizontal cell processes approached the terminal from either north or west (Fig. 2.11). At a fifth complex, one horizontal

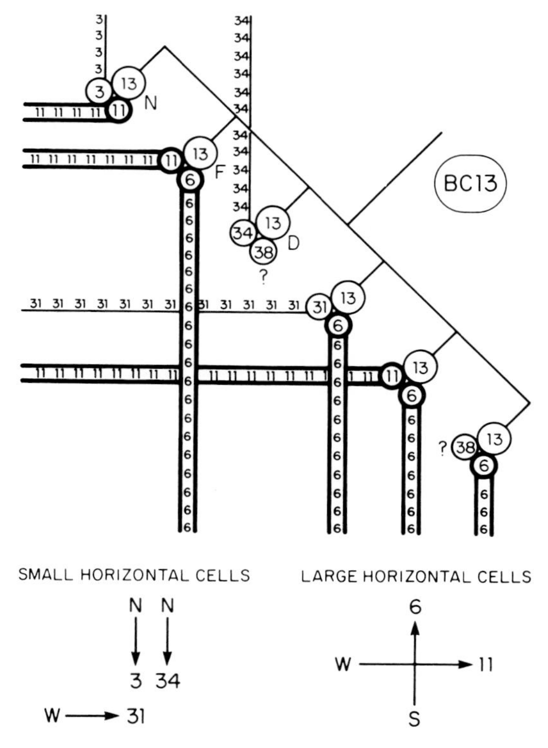

Fig. 2.12. Approach directions of horizontal cell processes connected to bipolar cell 13.

cell process approaching from west was paired with one, the approach direction of which could not be determined because it originated from a main horizontal cell process located outside the region contained in the series of sections. However, the end branch approached from the north.

Another bipolar cell (13) contacted four horizontal cells at six synaptic ribbon complexes. All horizontal cells approached the terminal from either south or west in the four complexes in which the approach directions of the horizontal cells could be determined (Fig. 2.12). A fifth complex, with approach directions from west and north, represented an exception. Figure 2.13 shows the approach direction of the horizontal cells connected to the other four bipolar cells that contributed endings to synaptic ribbon complexes. In the case of bipolar cell 8, which sent endings to four synaptic ribbon complexes, the horizontal cells approached from the same two perpendicular directions at all synaptic

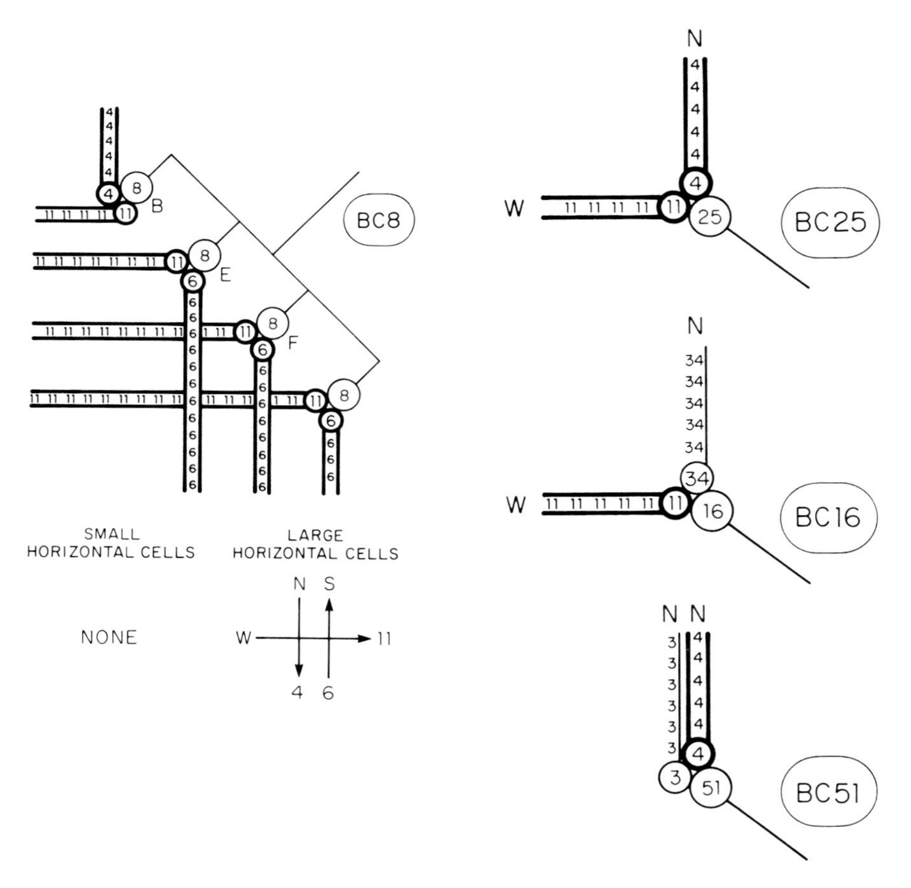

Fig. 2.13. Approach directions of horizontal cell processes connected to bipolar cells 8, 16, 25, and 51.

ribbon complexes. Of the other three bipolar cells only bipolar cell 51 offered an exception to this rule.

We can now summarize the observations reported so far that reveal that the neural connections are arranged according to certain systematic patterns.

1. Sclerad endings at synaptic ribbon complexes are always contributed by horizontal cells; vitread ending or endings are always contributed by bipolar cells.
2. The two horizontal cell endings are always contributed by two different horizontal cells.

3. These two horizontal cells approach the terminal predominantly from two perpendicular directions.
4. When several horizontal cells are involved in connections with one bipolar cell at several synaptic ribbon complexes, these horizontal cells approach the terminal from predominantly the same two perpendicular directions.

4. The Pairing of Two Types of Horizontal Cells

The double connections between a bipolar cell and horizontal cells could have a particular significance if the two horizontal cells transmit different types of information. The analysis revealed that in many synaptic ribbon complexes structural conditions for such a difference are fulfilled because two different types of horizontal cells are involved in the synaptic ribbon complexes. This conclusion was based on the observation that there are two distinct size classes of horizontal cell processes (Fig. 2.14). There are three size classes of neural processes from which end branches originate: the thick and the intermediate horizontal cell processes and the thin bipolar processes. These processes measured about 2, 0.8, and 0.25 μm, respectively.

The two types of horizontal cell processes are also distinguished by differences in their mutual contact relations. While the thick processes contact each other by broad contacts of the gap junction type, the medium-sized processes contact the thick processes through thin end branches. Synaptic vesicles are present *in the thick processes* at the site of contact.

The medium-sized processes were in some cases traced to the soma of the neuron, and the tracing then showed that they belonged to horizontal cells smaller than those contributing the large processes. They are thus not narrowed end branches of the latter processes. All these observations justify the conclusion that the large and medium-sized processes belong to two different types of horizontal cells. These horizontal cells likely correspond to the two types observed by Gallego (1971) in the rabbit retina in Golgi-stained preparations. Most of the basis for his identification of the two types of cells was the different size of the cells and the different thickness and the tapered shape of the processes of the small horizontal cells.

The two types of horizontal cells will be referred to as the large and the small horizontal cells. The approach directions of the two types of horizontal cells are shown in Fig. 2.10. Obviously, asymmetry in the approach directions of the small horizontal cells is even more pronounced than in those of the large horizontal cells. Small horizontal cells thus

Fig. 2.14. Two types of horizontal cells. (A) A typical process of a large horizontal cell. The small branch connects the process to terminal 1. (B) A typical process of a small horizontal cell contributing several small branches to terminal 1 (encircled area) and several end branches that end in contact with large horizontal cells. From Sjöstrand (1976).

approach only from north or west, with a considerable dominance for north (four out of five).

With two types of horizontal cells involved at the synaptic ribbon complexes, it is of interest whether their connections to bipolar cells are arranged according to some systematic, nonrandom pattern. First, the pair of horizontal cell endings at the synaptic ribbon complexes are contributed either by one small and one large horizontal cell or by two large horizontal cells. The small horizontal cells were never found paired, showing a restriction in the arrangement of the horizontal cell endings. In seven out of thirteen synaptic ribbon complexes, one large horizontal cell was paired with one small horizontal cell, while in the other six ribbon complexes two large horizontal cells were paired.

A systematic arrangement of the horizontal cell connections of the bipolar cells became obvious when the pattern of connections of individual bipolar cells was considered. Of the six bipolar cells contributing endings to synaptic ribbon complexes at terminal 1 four contacted complexes with endings contributed by one large and one small horizontal cell. Such synaptic ribbon complexes will be referred to as mixed complexes. One of the other two bipolar cells contacted horizontal cells at four synaptic ribbon complexes, and all contacts involved only large horizontal cells. The remaining bipolar cell contacted horizontal cells only at one synaptic ribbon complex.

Systematic patterns also emerged when the approach directions of the small horizontal cells were considered. One bipolar cell could contact several small horizontal cells at different synaptic ribbon complexes. In most cases all processes from the small horizontal cells approached the terminal from one and the same direction (Fig. 2.11).

Extensive analysis of the processes of small horizontal cells revealed that the processes do not contact photoreceptors located in a direction opposite that from which the processes approach the terminal. For instance, the four small horizontal cells contacted by bipolar cell 1 at synaptic ribbon complexes at terminals 1 and 2 approached these terminals from the north but did not have any connections to photoreceptors south of these terminals. It seems highly significant that this pattern was the same for all four small horizontal cells.

This means that these horizontal cell processes will be influenced directly only by the illumination of an area north of the terminals because they receive input only from photoreceptors within this area (Fig. 2.15). However, as mentioned above, the analysis also showed that end branches from these processes contact large horizontal cells and that synaptic vesicles are present in the large horizontal cells at the site of

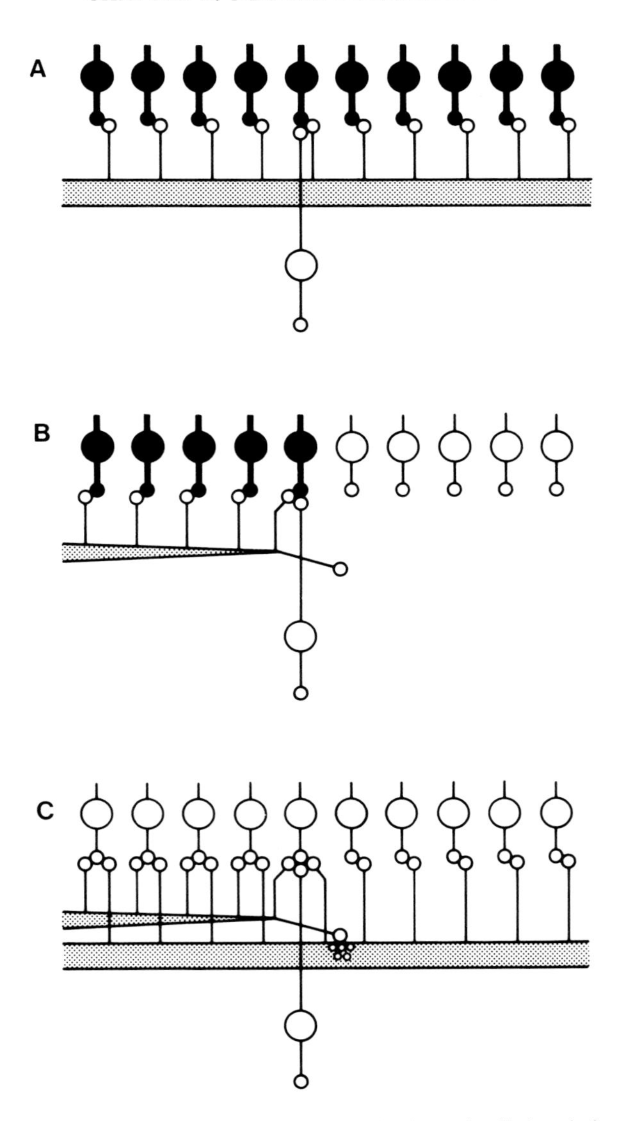

Fig. 2.15. Connections of large (A) and small (B) horizontal cells in relation to terminal 1, illustrating schematically the difference in the areas within which photoreceptors can influence the two types of horizontal cells. (C) The connection between large and small horizontal cells. Note the rather uniform thickness of the large horizontal cell processes and the tapered shape of the small horizontal cell processes.

contact. The conditions for the large horizontal cells influencing the small horizontal cells through these connections are therefore fulfilled.

We can now add the following features supporting the systematic, nonrandom nature of horizontal cell-bipolar cell connections:

5. At synaptic ribbon complexes, either endings from two types of horizontal cells, the large and the small, are paired or the endings of large horizontal cells are paired. The endings of small horizontal cells are never paired.
6. Most bipolar cell endings contact mixed synaptic ribbon complexes.
7. Bipolar cells that contact mixed synaptic ribbon complexes also contact complexes in which both horizontal cell endings are contributed by large horizontal cells. At the latter complexes, the two large horizontal cells always approach the terminal from perpendicular directions (Figs. 2.11A, 2.12, and 2.13).
8. Small horizontal cells that are contacted by one bipolar cell approach the terminal mostly from the same direction. They never approach from opposite directions.
9. The small horizontal cell processes make no contacts with photoreceptors located on the side of the terminal opposite from which they approach the terminal. For example, if the processes approach the terminal from the north, they make no connections with photoreceptors south of the terminal.

Already at this stage of our analysis we feel confident that the neurons in the outer plexiform layer are connected according to certain well-defined patterns and that we are not dealing with a random neural network. Again, it should be pointed out that this order was discovered *after* the three-dimensional reconstruction had been finished and could not have been anticipated during the reconstruction. No preconceived concept could consequently have influenced the reconstruction. Errors in the reconstruction are, from a statistical point of view, more likely to contribute randomness than order.

5. The Synaptic Ribbon Complex in Rod Terminals

In contrast to the cone terminals, only one synaptic ribbon complex occurs in rod terminals in the retinas of rabbits and guinea pigs. Unfortunately, the possibility that bipolar cells and horizontal cells are connected according to particular patterns has not been analyzed in rod terminals. Such an analysis requires that the dendrites from individual bipolar cells be traced to many rod terminals to reveal the connections at many synaptic ribbon complexes.

There are considerable differences between rod and cone terminals in the structure of the synaptic ribbon complexes in the rabbit retina. One basic difference is the size of the bipolar cell endings that contact the rods and the horizontal cells. In the cone terminals the bipolar cell endings to synaptic ribbon complexes consist of very thin end branches or thin side branches extending from a main end branch. In the rod terminals the bipolar cells contribute larger endings to the synaptic ribbon complexes. The end processes in these terminals are thin as they pass through the most vitread part of the invagination of the rod membrane, but then they widen to form bulbous, clublike endings (Figs. 2.16–2.18). The end branches of the horizontal cells are thin where they enter the invagination and remain thin as they pass by the bipolar cell endings. Sclerad to these endings the horizontal cell end branches widen to form large, irregularly shaped endings with many projections (Figs. 2.16–2.18).

In contrast to the synaptic ribbon complexes in cone terminals, the two horizontal cell endings in rod terminals are separated only partially by the septum containing the synaptic ribbon. Vitread to the septum the two endings are in contact within a narrow zone. This is the only contact between the two horizontal cell endings at the synaptic ribbon complexes. Thus, in the vitread part of the complex the thin parts of the horizontal cell endings are not in contact, in contrast to the situation at cone terminals. In the rod terminals in the part of the rabbit retina analyzed, as a rule two bipolar cells were connected to each synaptic ribbon complex.

The contact at the synaptic ribbon complexes between the bipolar cells and the horizontal cells occur in three different ways. According to a first pattern the bulbous bipolar cell ending is in contact with the broad vitread surface of both horizontal cell endings symmetrically relative to the synaptic ribbon (Figs. 2.16, 2.17). These contacts involve large areas, and the size of the bipolar cell ending appears to be adjusted to allow extensive contacts between the bipolar cell and both horizontal cells.

A second pattern is characterized by the bipolar cell contacting the rod terminal with two end branches, one contacting the surface of the rod terminal at its vitread pole, and the other making a brief triad type of connection to the two horizontal cells at the synaptic ribbon complex. The large size of the latter bipolar cell ending has no relationship to the size of the areas of contact with the horizontal cells. Instead, the large size allows an extensive contact between the bipolar cell and the rod membrane (Fig. 2.17.B, C).

A third pattern is entirely different. The large bipolar cell ending is located in a separate branch of the invagination that extends laterally from

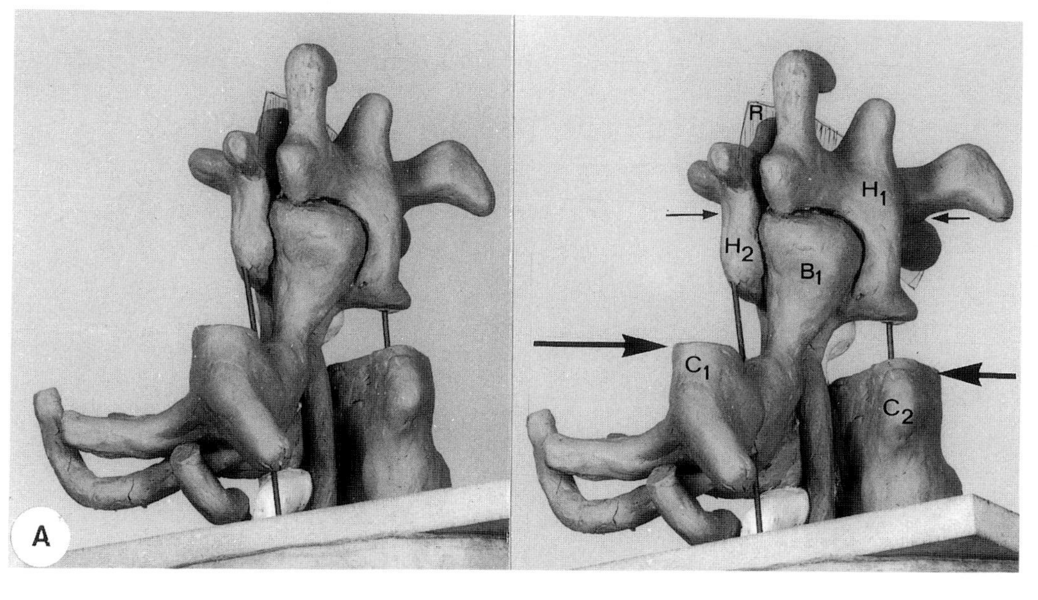

Fig. 2.16.A. Stereophotographs of three-dimensional reconstruction of the synaptic ribbon complex in the rod terminal marked Y in Fig. 1.20. Sclerad direction upward. Bipolar cell B_1 is connected to horizontal cells H_1 and H_2 over large areas. Two cone processes (C_1 and C_2) contact the vitread surface of the rod terminal. Large arrows in Figs. 2.16 and 2.17 indicate the level to which the vitread pole of the terminal reaches. The parts of the model located above this level consist of invaginated endings, whereas the parts below this level are located vitread to the terminal in what will be referred to as the subsynaptic neuropil in Chapter 4.1. The small arrows indicate the boundary between the vitread and the sclerad parts of the synaptic ribbon complex. R, the synaptic ribbon.

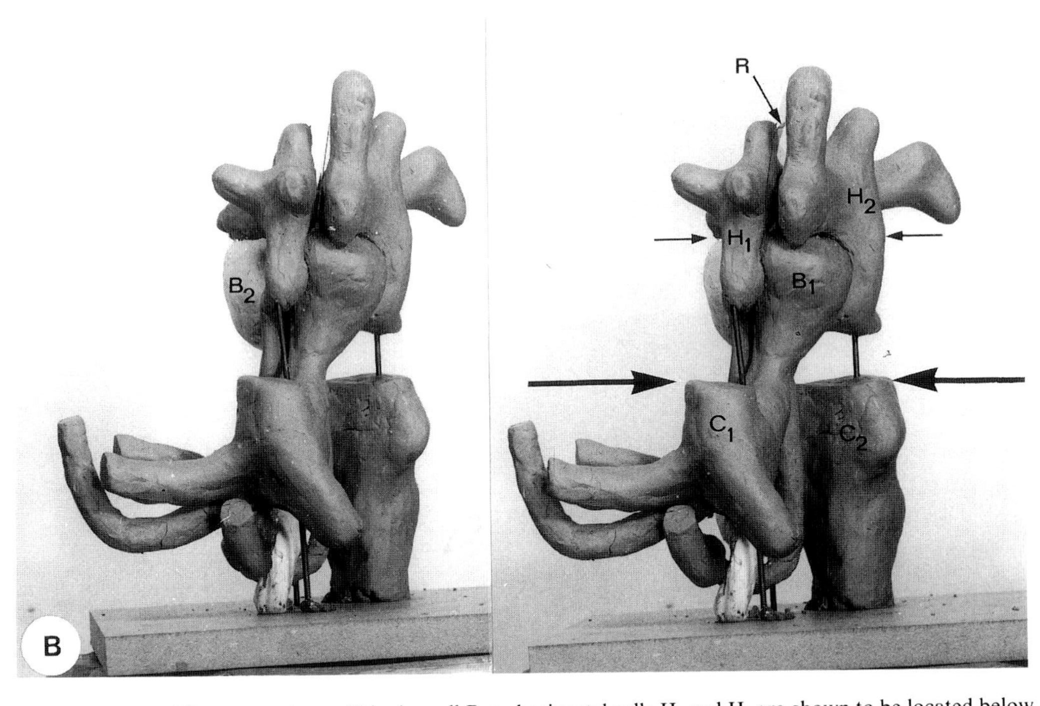

Fig. 2.16.B. The connections of bipolar cell B_1 to horizontal cells H_1 and H_2 are shown to be located below the vitread edge of the synaptic ribbon, which is characteristic of a typical triad type of connection.

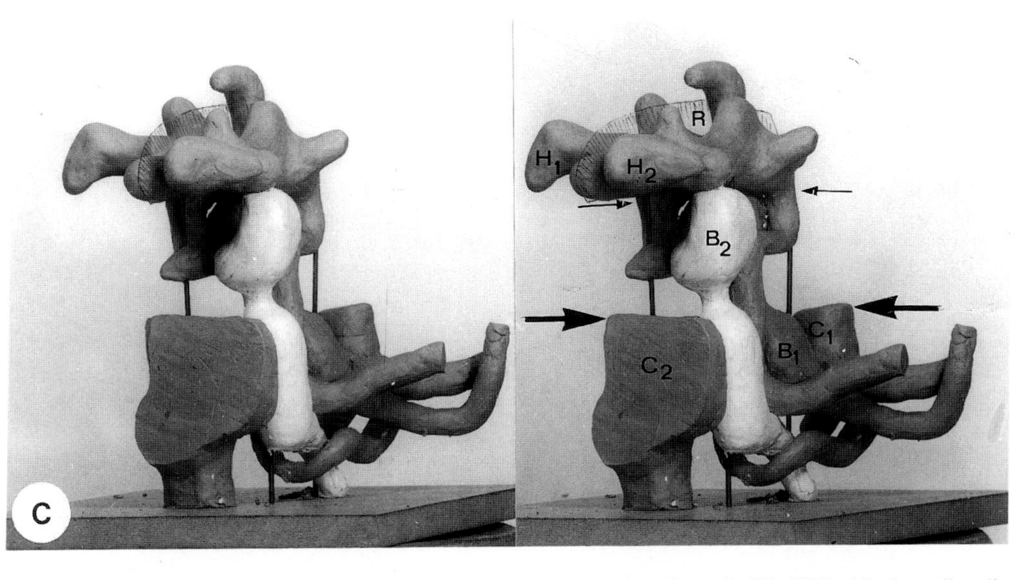

Fig. 2.16.C. The model as viewed from a direction opposite to that shown in (A). White bipolar cell ending B_2 is connected to horizontal cell H_2, laterally away from the synaptic ribbon. In addition to this connection in the sclerad part of the complex, bipolar cell B_2 is connected to the thin part of both horizontal cells H_1 and H_2 in the vitread part. These contacts involve small areas. Most of the surface of the bipolar cell B_2 ending is in contact with the rod membrane. The shape and size of the large bulbous ending have no relationship to the size of the bipolar cell–horizontal cell contact areas.

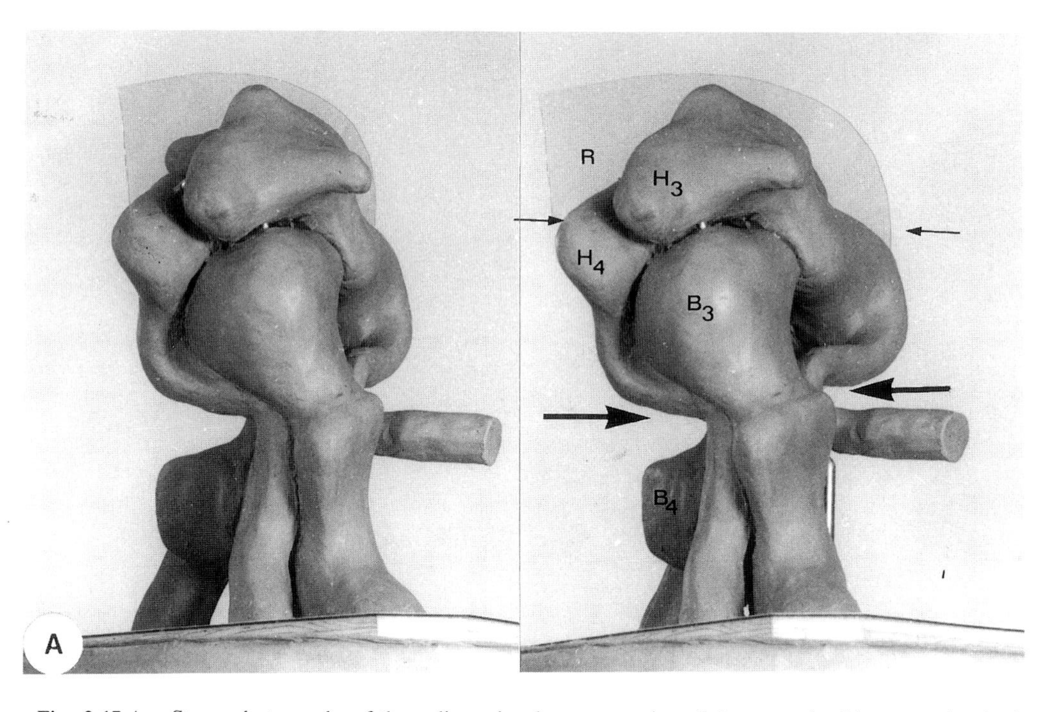

Fig. 2.17.A. Stereophotographs of three-dimensional reconstruction of the synaptic ribbon complex in the rod terminal marked X in Fig. 1.20. Bulbous bipolar cell ending B_3 makes a typical triad connection to the sclerad parts of the horizontal cell endings H_3 and H_4.

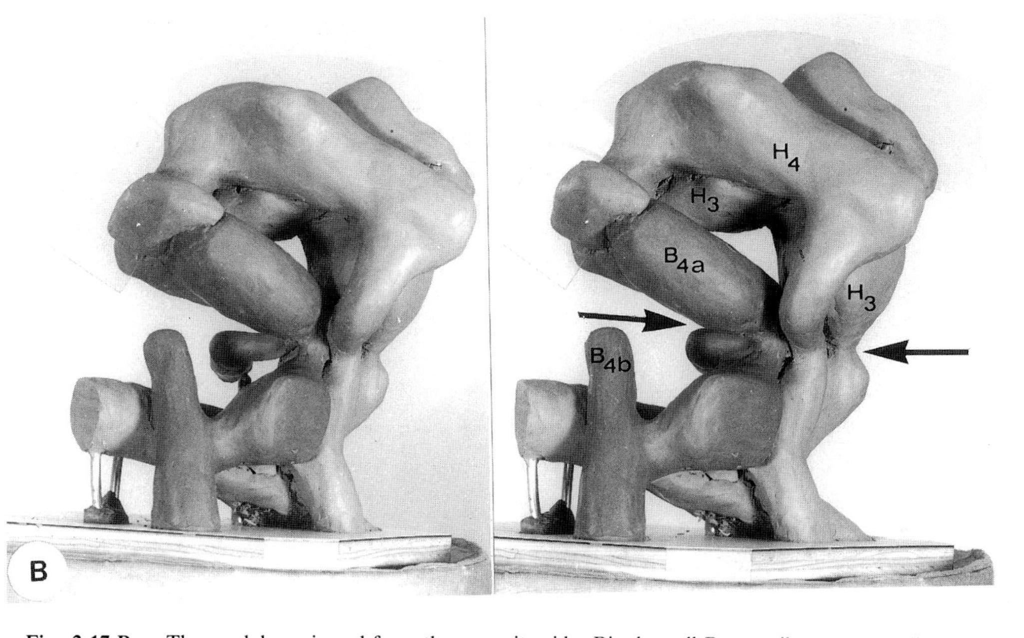

Fig. 2.17.B. The model as viewed from the opposite side. Bipolar cell B_4 contributes two endings, one of which (B_{4a}) is connected to the sclerad parts of horizontal cells H_3 and H_4 in an asymmetric way, and with the connections involving small contact areas. Most of the large bulbous bipolar cell ending exposes a large surface area that is in contact with the rod membrane. A second ending of bipolar cell B_4 (B_{4b}) is connected to the vitread surface of the rod terminal.

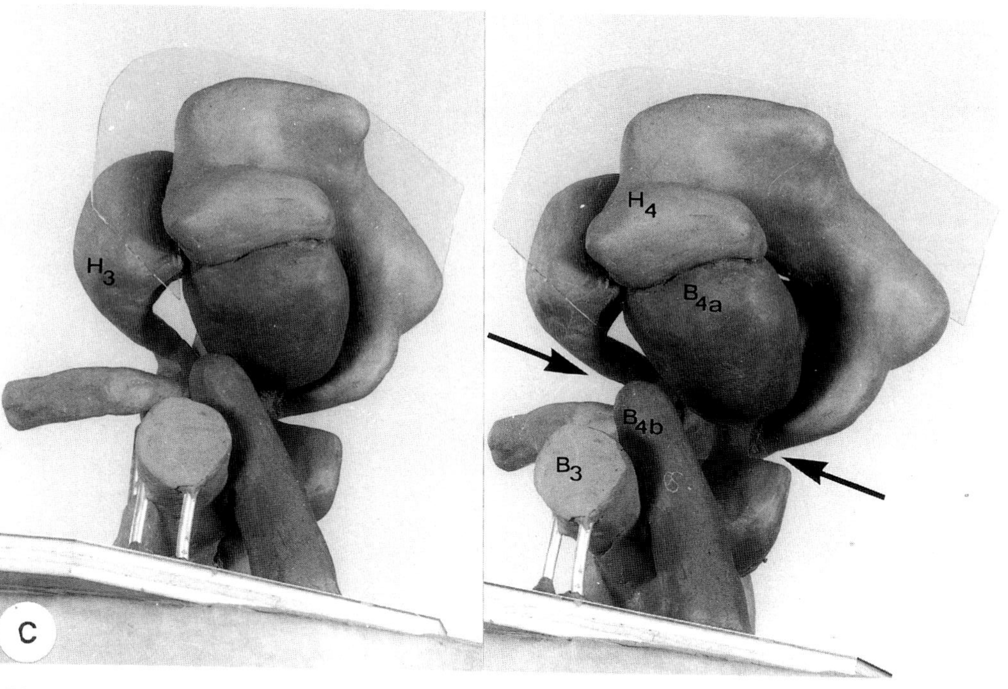

Fig. 2.17.C. The asymmetric triad type of connection of bipolar cell B_4 involves mainly horizontal cell H_4 and the areas of contact are considerably smaller than the connections of bipolar cell B_3 to the two horizontal cells.

the branch containing the synaptic ribbon complex. The ending contacts the widened sclerad part of only one horizontal cell ending at a site located far laterally from the center of the synaptic ribbon complex. The connection to the horizontal cell is brief and involves a lateral extension of the horizontal cell ending, reached through a communication between the two separate invaginations. The bipolar cell also is in contact with the thin part of both horizontal cell endings in the vitread part of the synaptic ribbon complex. These contacts are topographically widely separated (Figs. 2.16.C, 2.18.D). The total surface area of contact between the bipolar cell and the horizontal cell is very small in comparison to the area over which the bipolar cell ending contacts the rod membrane.

We interpret these differences in the connections between bipolar cells, horizontal cells, and the rods as reflecting differences in the mode of communication between the neurons. The first pattern favors a strong influence of the horizontal cells on the bipolar cell, while the third pattern favors the influence of the rod over the horizontal cells. The second pattern appears to represent a combination of the other two patterns because it fulfills conditions for strong influence by both the rod and the horizontal cells. The third pattern of connections has certain similarities to the connections of bipolar cells described below that make lateral connections at the synaptic ribbon complexes in the cone terminals.

Characteristically, the two bipolar cell endings at a rod terminal are connected differently. The connection of one ending usually follows the first pattern with extensive contacts of the triad type. The other bipolar cell ending is connected to the rod and the horizontal cells according to either the second or the third pattern. Assuming that these patterns allow the transmission of different information, we would expect that the two bipolar cells connected to one rod transmit different types of information. This would give significance to the fact that in the rabbit retina the rod terminals are contacted by two bipolar cells.

The tracing of the horizontal cell end branches to rod terminals showed that these branches could be contributed either by small or by large horizontal cells as in cone terminals.

In the guinea pig retina the first pattern of connections dominated and applied to both bipolar cells when the terminal was contacted by two bipolar cells. There were also other differences in the structure of the synaptic ribbon complexes in the guinea pig retina. The number of neurons contributing endings to these complexes thus could vary. Three types of synaptic ribbon complexes could be distinguished on this basis (Sjöstrand, 1965).

The first structural pattern of the synaptic ribbon complex consisted of a pair of horizontal cell endings that were both contacted by a pair of

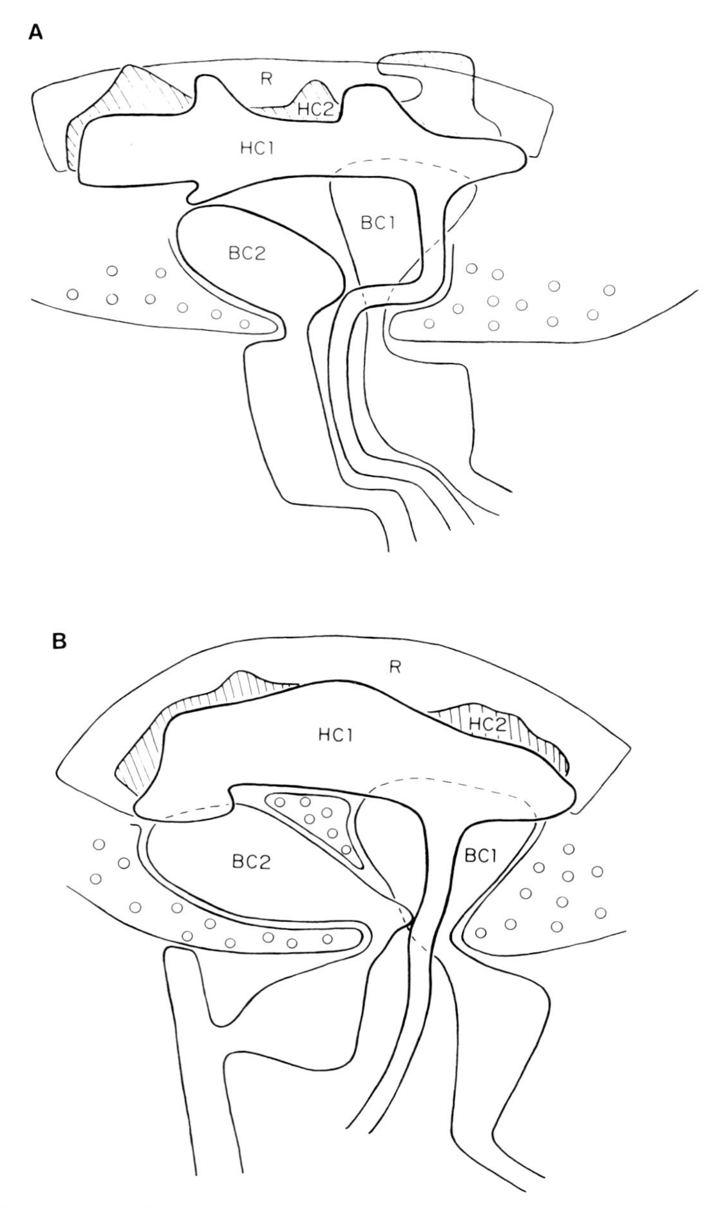

Fig. 2.18. Schematic drawings of the three types of bipolar cell connections at rod terminals. (A) Triad type of connection of bipolar cell BC1 and lateral connection to the horizontal cells of bipolar cell BC2. (B) Triad type of connections of both bipolar cells. Bipolar cell BC2 also contacts the surface of the rod terminal with a special ending,

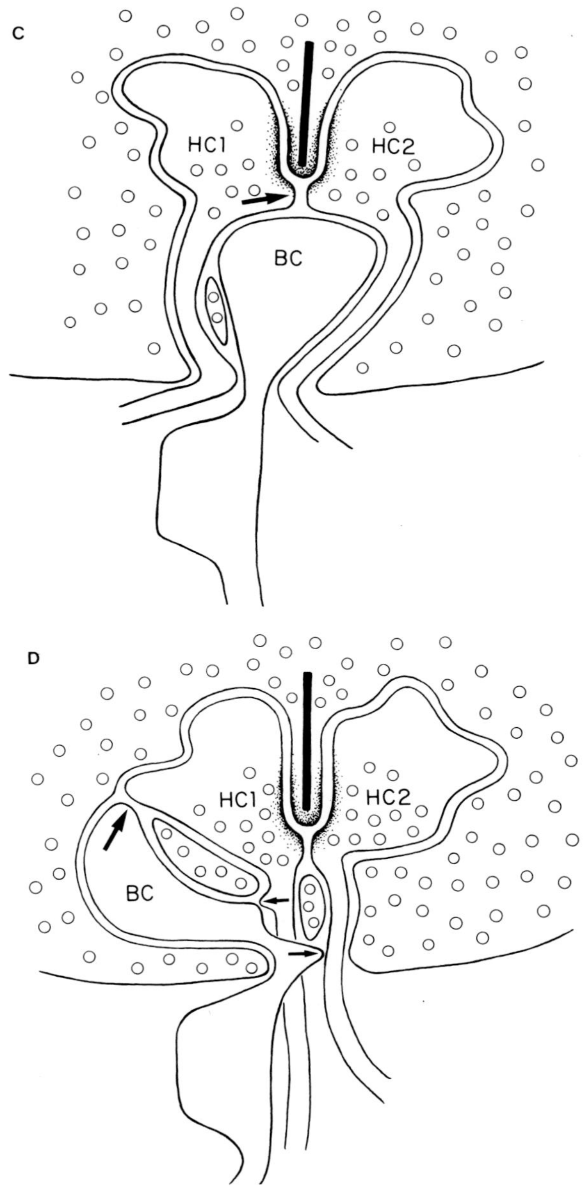

corresponding to the connections illustrated in Fig. 2.17.B and C. (C) A typical triad type of connection in synaptic ribbon complex viewed end on. Arrow points connections between the sclerad parts of the two horizontal cell endings. (D) Lateral connection of bipolar cell such as the connection of bipolar cell B_2 in Fig. 2.16C. The lateral connection to the sclerad part of horizontal cell ending HC1 (thick arrow) is combined with brief connections to both horizontal cell endings in the vitread part of the complex (small arrows).

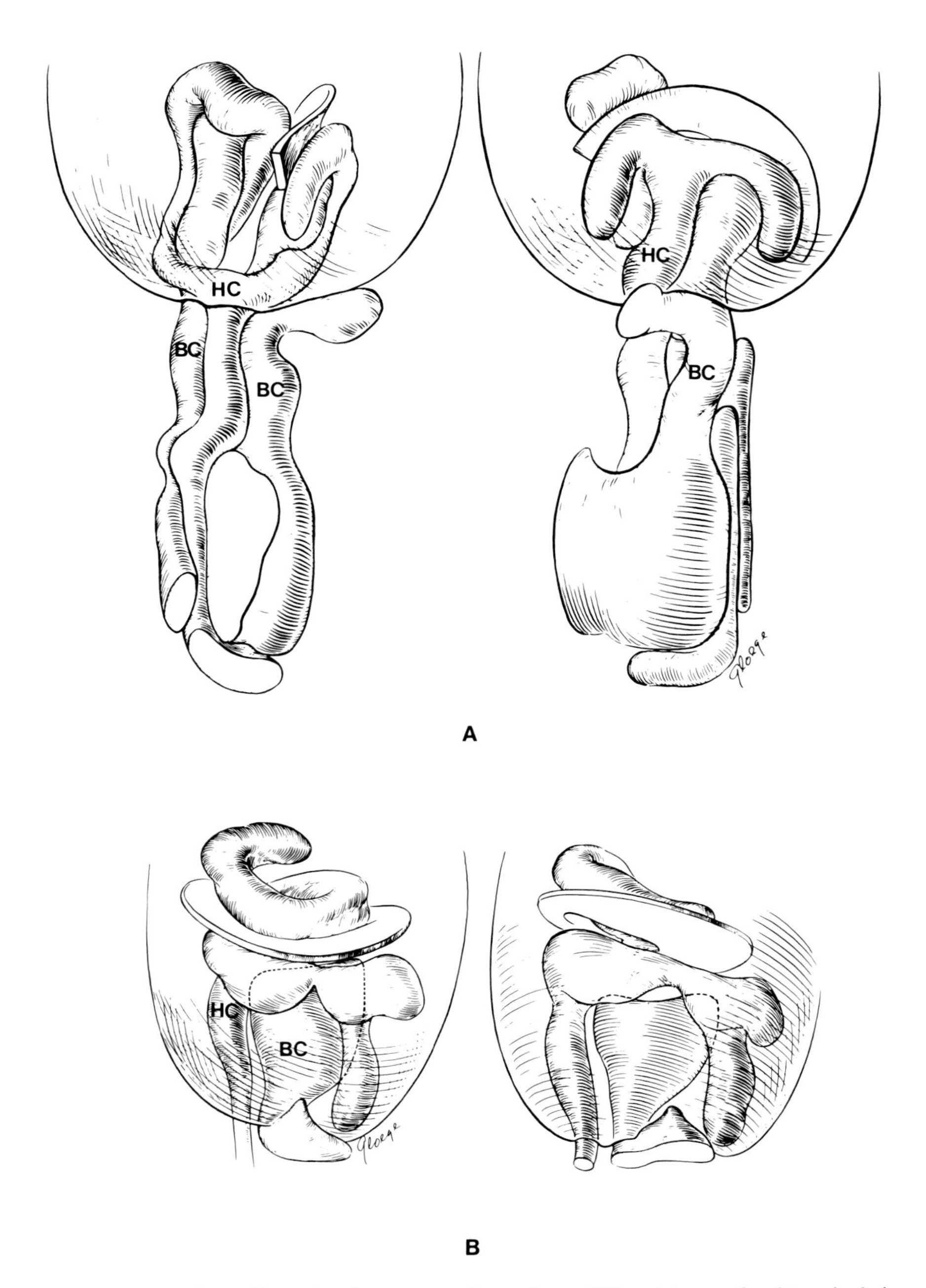

Fig. 2.19. Three-dimensional reconstructions of two different types of rod terminals in the guinea pig retina. (A) The endings of two bipolar cells make individual contacts with two branches of the same horizontal cell ending. (B) One single bipolar cell ending contacts a single horizontal cell ending that is coiled about one and a half turns, with the synaptic ribbon located between the turns.

bipolar cell endings. According to a second pattern, a pair of bipolar cell endings contacted a single branched horizontal cell ending. A single bipolar cell ending contacting a single horizontal cell ending represented the third pattern. Two of these patterns are illustrated in Fig. 2.19. They were observed repeatedly in the guinea pig retina.

When a single horizontal cell ending was involved, it could either split into two end branches or the ending could turn to form a spiral. The synaptic ribbon is located between the two end branches in the first case and between the turns in the one-and-a-half-turn spiral in the second case. In both cases, one edge of the septum containing the ribbon extends close to the contact region between the horizontal cell and the bipolar cells. The septum thus partially separates the two end branches or the turns of the spiral of the horizontal cell endings.

6. The Efferent Bipolar Cell

Ramon y Cajal (1892, 1933) and Polyak (1941) identified certain neurons that extend between the inner and the outer plexiform layers that they classified as centrifugal bipolar cells because their branching pattern in the inner plexiform layer resembles a dendritic type while their branching pattern in the outer plexiform layer resembles an axonal type. These cells are now usually referred to as interplexiform cells (Gallego, 1971) and they have been observed in a variety of species. Their synaptic connections in the outer plexiform layer are unknown.

The three-dimensional reconstruction of the two cone terminals revealed the presence at both terminals of large invaginated bipolar cell endings characterized by a dense accumulation of synaptic vesicles and by a rather specialized structure of the neural membrane. These features are illustrated in Figs. 1.23, 1.24, and 1.32.

At terminal 1 the single neuron characterized by such endings branched to contribute three large endings that were deeply invaginated. All endings were filled with synaptic vesicles. This neuron was traced to three other cone terminals and in all terminals it contributed invaginated endings densely filled with synaptic vesicles. It is therefore justifiable to conclude that this type of ending characterizes this neuron. It distinguished it from all the other neurons contacting the terminals. One neuron (15_2) with the same synaptic vesicle-filled invaginated endings was present also at terminal 2 (Fig. 2.20). Such endings are therefore present generally in the cone terminals. They have not been observed in rod terminals.

The evaluation of these neurons as bipolar cells was based on the fact that their end branches originated from a single thin process that was traced to the inner nuclear layer.

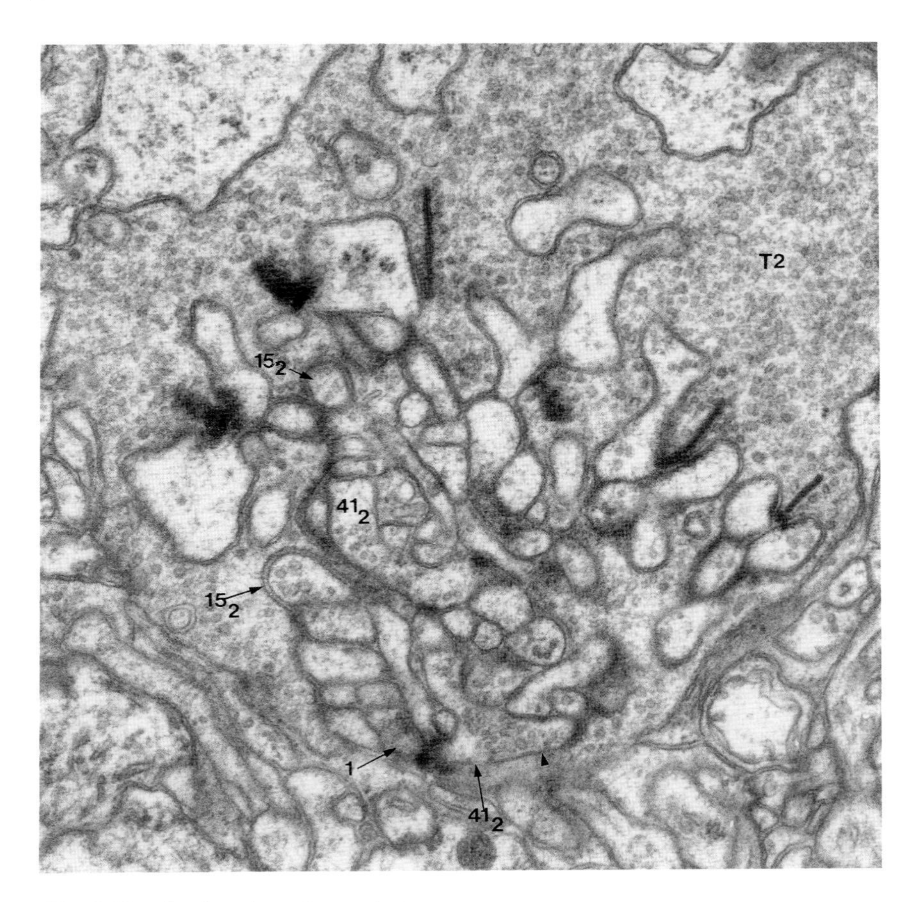

Fig. 2.20. Section through terminal 2 (T2) and through neural endings located just vitread to the terminal (in the subsynaptic neuropil). Bipolar cell 15_2 contacts the terminal with endings containing synaptic vesicles in the same way bipolar cell 14 contacts terminal 1. Synaptic vesicles are also present in bipolar cell endings located vitread to the terminal as shown by the endings of bipolar cell 41_2. To distinguish the neural end processes contacting terminal 2 from those contacting terminal 1, the numbers of the former processes are indicated by the subscript 2. In the lower part of the picture bipolar cell process 41_2 extends an end branch to the left that contacts a process of bipolar cell 1. There are no synaptic vesicles in the bipolar cell 41_2 process at its contact to bipolar cell 1 (arrow), whereas there are numerous synaptic vesicles in the end branch of this bipolar cell extending to the right (arrowhead). In adjacent sections the end process of bipolar cell 1 contacting bipolar cell 41_2 in this area contained numerous synaptic vesicles that are not shown here because the end process of bipolar cell 41_2 was partially superimposed on the end process of bipolar cell 1 and because the section cuts tangentially through the neural membrane of the bipolar cell 1 end process. 30,000×. From Sjöstrand (1978).

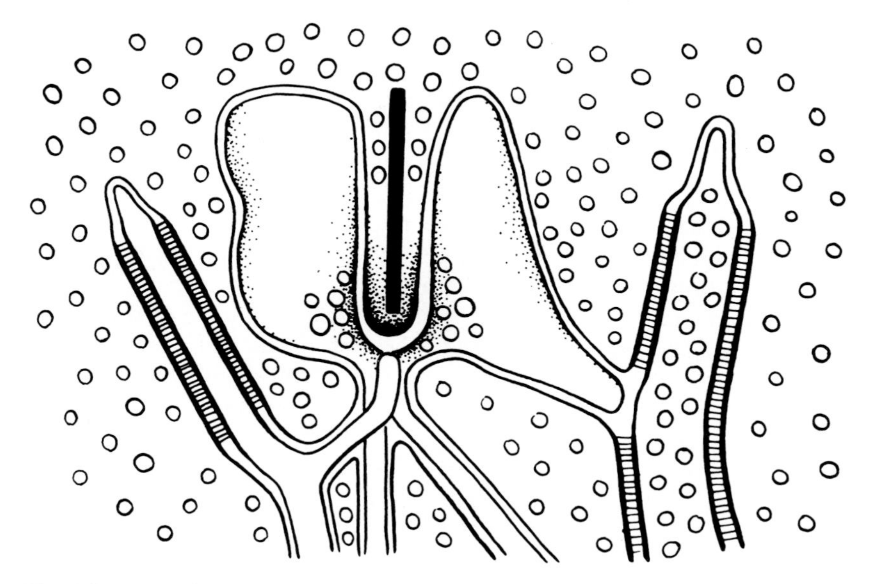

Fig. 2.21. Drawing showing a lateral connection between an efferent bipolar cell and a horizontal cell ending at a synaptic ribbon complex.

The processes of these neurons in the outer plexiform layer definitely contributed predominantly presynaptic endings, as judged from the presence of synaptic vesicles in the endings. Since the analysis did not involve a complete reconstruction of the neurons, their input remains unknown. It is tentatively assumed, however, that connections in the inner plexiform layer contribute the input, and that these neurons correspond to the centrifugal bipolar cells of Ramon y Cajal and Polyak.

Until the relationship of these neurons to the interplexiform cells has been established by more extensive three-dimensional analysis, they will be referred to as efferent bipolar cells without any claim that they are a type of neuron different from those already described by neuroanatomists.

The efferent bipolar cells make limited contacts with horizontal cells. At terminal 1 the efferent bipolar cell contacted only one of the two horizontal cell endings at four synaptic ribbon complexes. The contact involves a lateral extension of the horizontal cell ending, and the contact surface is located laterally from the ribbon, a lateral contact (Figs. 1.20, 2.21). The contact was made by the bipolar cell endings when they passed by the synaptic ribbon complexes to extend sclerad considerably deeper in the terminal. This type of bipolar cell–horizontal cell contact characteristically involves only one of the two horizontal cells at the synaptic ribbon complexes and only small horizontal cells are contacted by this bipolar cell at different synaptic ribbon complexes.

7. The Core Bipolar Cell

In both cone terminals that were reconstructed, one neuron of the bipolar cell type differed conspicuously from all other neurons by its end branch being positioned centrally among the neural processes vitread to the terminals and by an elaborate system of side branches that contacted other neural processes in this region (Process 39 in Figs. 1.19–1.21.) Accumulations of synaptic vesicles were present at most of these side branches at the site of contact. Five out of ten bipolar cells contributing processes to terminal 1 were contacted by this neuron, and the same pattern of extensive bipolar cell contacts applied to the corresponding bipolar cell at terminal 2 (Process 41_2 in Fig. 2.20).

The central position of these processes among the neural processes vitread to the two cone terminals, the relatively large volume they occupied, and the elaborate system of branches contacting other processes (Fig. 2.22) justify designation of these neurons as a special type. They are referred to as *core bipolar cells* because of their location and the fact that they show all structural characteristics of a bipolar cell. The central location enables a core bipolar cell's process to contact a large number of other bipolar cell processes with short branches.

The contacts made by branches from the core bipolar cell processes do not involve all surrounding processes indiscriminately (Fig. 2.23). At terminal 1 there was a total of nine contacts with one particular bipolar cell process (bipolar cell 13). This latter process split into six branches that contacted the terminal. All these end branches were contacted individually by the core bipolar cell. In addition to these six end branches, bipolar cell 13 contributed two special end branches that contacted the core bipolar cell exclusively vitread to the terminal, clearly revealing a special relationship between these two bipolar cells.

Five branches extending from the core bipolar cell contacted three other bipolar cells. Three of these branches involved the efferent bipolar cell (Fig. 2.23). A fifth bipolar cell contacted the core bipolar cell with a special end branch. There were no contacts between the core bipolar cell and horizontal cells vitread to the terminal.

The core bipolar cell contacted the terminal with invaginated endings. No special branches extended to any synaptic ribbon complexes. The end processes did, however, pass by one of the horizontal cell endings at several synaptic ribbon complexes to make a lateral contact.

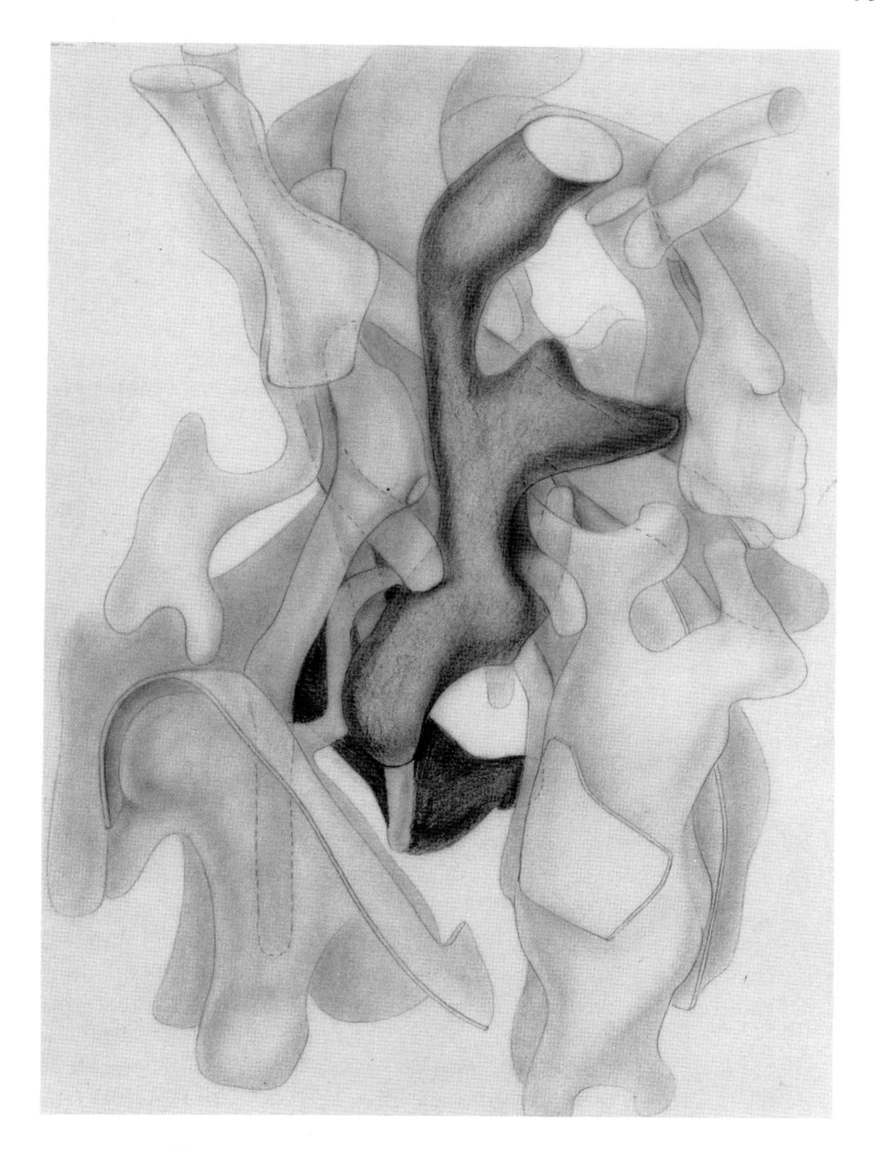

Fig. 2.22. The core bipolar cell at terminal 1 (shaded).

The dendrite of the core bipolar cell extending to terminal 1 contacted two rod terminals located adjacent to this terminal. It sent off one short branch to each of these rod terminals and each branch split into two end

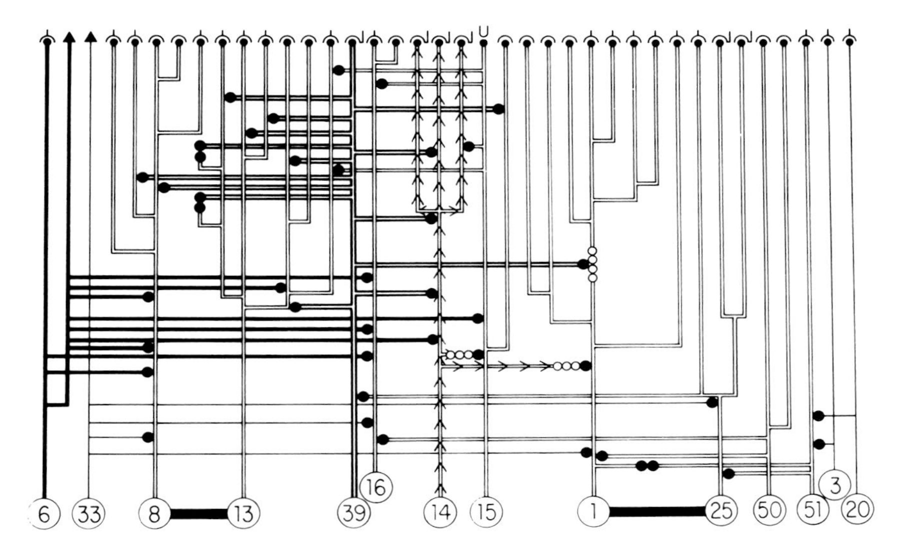

Fig. 2.23. Account of neural connections just vitread to terminal 1 (in the subsynaptic neuropil) with the connections of the core bipolar cell 39 emphasized. Double lines, bipolar cell processes; single lines, horizontal cell processes; thick double lines, the core bipolar cell process; arrowed double lines, the efferent bipolar cell process. The processes of bipolar cells 8 and 13 are connected, as are the processes of bipolar cells 1 and 25. These two pairs of bipolar cells were associated particularly closely with a narrow separation of the neural membranes that were particularly thin at the area of contact of the two processes of each pair, as illustrated in Fig. 1.33. Redrawn from Sjöstrand (1976).

branches. One of these end branches ended at processes vitread of the rod terminal without contacting the terminal, while the other process made contacts with the surface of the rod terminal at its vitread pole. This pattern is in principle similar to the pattern of contacts at terminal 1.

The functional significance of the core bipolar cells can not be deciphered from the observed pattern of connections with photoreceptors and other neurons because the reconstruction did not include a large enough region. This neuron does, however, represent another example of a systematic pattern of neural connections.

The widespread occurrence of synaptic vesicles in the end branches that contacted other bipolar cells suggests that the core bipolar cell exerts a modulating influence on several bipolar cells. It is, however, not possible to guess what kind of influence this neuron may have since the input into the core bipolar cell is unknown. It is thus unknown whether this is an efferent bipolar cell representing part of a feedback circuit from the inner plexiform layer or whether it is an afferent bipolar cell that

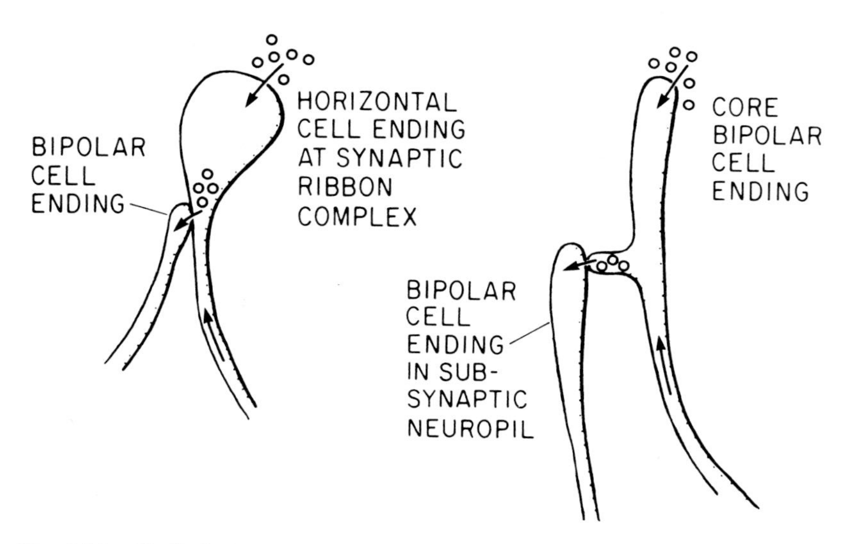

Fig. 2.24. Similarity in the arrangement of the endings of the horizontal cells at synaptic ribbon complexes and the endings of the core bipolar cell. In both cases the neurons are connected to bipolar cells with synaptic vesicles in the core bipolar cells and in the horizontal cells at the endings, indicating that they are presynaptic components, while the terminal endings are connected to the photoreceptor which is here the presynaptic component. The difference is that the presynaptic connections of the horizontal cells are located at the synaptic ribbons, whereas the corresponding connections of the core bipolar cell are located vitread to the terminal (in the subsynaptic neuropil).

modulates centripetal transmission of information by other bipolar cells. In the first case, the connection to the terminal would have a modulating effect on the feedback signal, while in the second case, these connections would represent the real input to the neuron.

Assuming that, like an interplexiform cell, the core bipolar cell transmits information over a long distance, Sjöstrand (1974, 1978) proposed that it is efferent relative to certain other bipolar cells but not with respect to the photoreceptor. This arrangement has a certain similarity to that of the horizontal cells at the synaptic ribbon complexes. These horizontal cells thus receive along a main process an input from photoreceptors that modulates their influence on the bipolar cell at the ribbon complex, while the endings at the complex allow the photoreceptor to modulate this influence locally. The similarity in the arrangement of the horizontal cells at synaptic ribbon complexes and of the core bipolar cell is illustrated in Fig. 2.24.

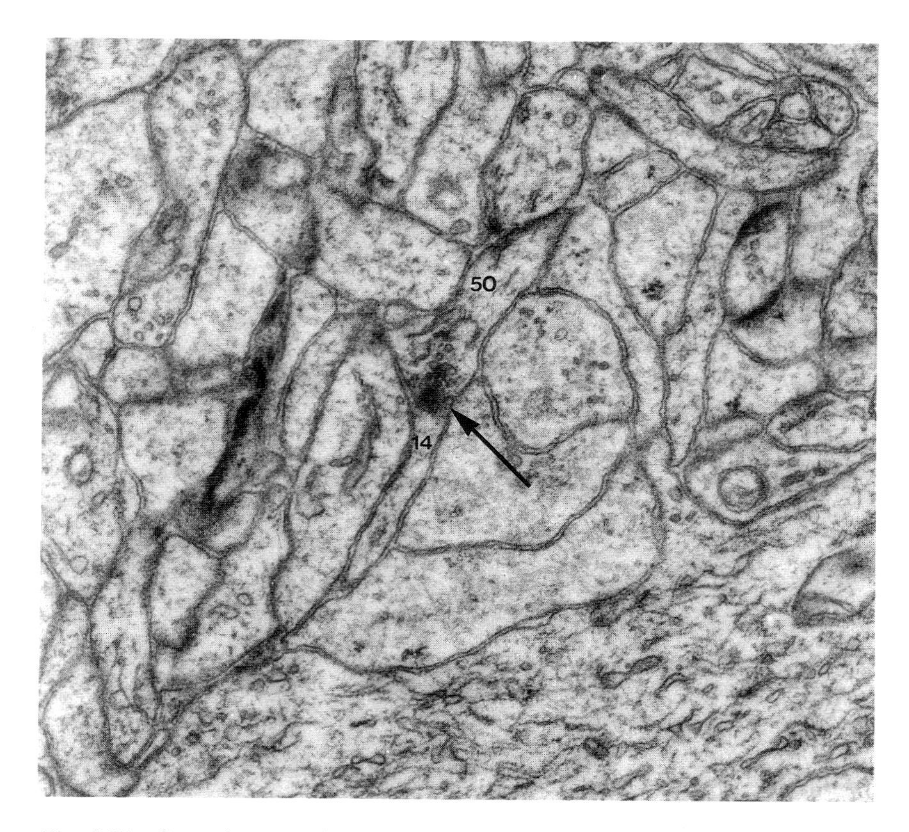

Fig. 2.25. Synaptic connection between bipolar cells in the common neuropil (arrow). The connection involves bipolar cells 50 and 14, and synaptic vesicles are present in the process of bipolar cell 50. Bipolar cell 14 is the efferent bipolar cell to terminal 1. 30,000×.

8. Bipolar Cells Not Connected to Horizontal Cells at the Terminal

Two bipolar cells did not contribute any end branches to synaptic ribbon complexes, and neither did they contact any horizontal cell endings outside these complexes. From one of these bipolar cells one additional end branch was traced to another cone terminal, and the contact pattern was the same as that at terminal 1. It involved two endings, one of which contacted one other bipolar cell just vitread to the terminal, while the other ended in a shallow invagination of the cone membrane. This bipolar cell also made synaptic contact with another bipolar cell at some distance away from the terminal. The synaptic nature of the contact between these two bipolar cell processes was shown by the specialized structure of the plasma membranes at the site of contact and by the accumulation of synaptic vesicles in one of the processes (Fig. 2.25).

One bipolar cell of this type was connected to two small horizontal cells through special branches extending from the horizontal cell processes. These contacts were located just vitread to terminal 1. Both horizontal cells approached the terminal from north. Neither of these two bipolar cells contacted any large horizontal cells outside the terminal.

We conclude that there are considerable differences among the neural connections of bipolar cells at cone terminals. The bipolar cells must therefore contribute in very different ways to the function of the retina.

Electron microscopic analysis has revealed that the class of neurons grouped as bipolar cells on the basis of the morphology shown by light microscopy consists of a functionally very heterogeneous collection of neurons that can be distinguished by their neural connections.

Chapter 3: The Functional Significance of the Observed Neural Connections

1. The Basis for Deducing the Responses of Bipolar Cells to Light Stimuli

The considerable variation in the way the bipolar cells are connected to photoreceptors, to horizontal cells, and to other bipolar cells must indicate that bipolar cells respond in several different ways to the input from photoreceptors. If we combine the structural information with information contributed by electrophysiological recordings from retinal neurons, we can deduce the most likely responses of the bipolar cells to the input from the photoreceptors. It is possible to deduce not only the sign of the bipolar cell response but also the pattern according to which the potential difference across the bipolar cell membrane varies when the light stimulus changes. However, these deductions require that certain basic aspects of the physiology of retinal neurons as revealed by electrophysiological recordings be included. Let us therefore now review these aspects of retinal physiology.

All neurons in the outer plexiform layer respond to a stimulus by a graded change in the state of polarization of the neural membrane, which can be recorded as a graded change in the electric potential difference across the membrane. The responses therefore differ from the all-or-none responses of spike-generating neurons, which transmit information over distances longer than those involved here.

The imposed changes can be an increase in the electric potential difference, pushing the membrane toward hyperpolarization, or a decrease in the potential difference, pushing the membrane toward depolarization. When the change is large enough, a locally imposed change in the state of polarization of the membrane will be conducted along the neural process. When the membrane is pushed far enough toward depolarization a transmitter is released in a graded fashion at the synaptic connections of the neural process.

The dominant mechanism of transmission of signals between the neurons in the outer plexiform layer is chemical. Gap junctions have been observed connecting the processes of the large horizontal cells (Yamada and Ishikawa, 1965; O'Daly, 1967; Stell, 1967; Sjöstrand, 1974), and it has been shown that electrophysiologically these connections are low-resistance electrical synapses (Kaneko, 1971).

Structural analysis has revealed synaptic vesicles at many contacts between the retinal neurons, agreeing with predominantly chemically mediated transmission between these neurons. In the case of the photoreceptors, electrophysiology supports a chemical transmission involving the release of a depolarizing transmitter (Trifinov, 1968; Byzov, 1966; Trifinov and Ostrovskii, 1970; Baylor *et al.*, 1971). This transmitter is released continuously in the dark, and its release is reduced when the receptor is stimulated by light. Horizontal cells release a hyperpolarizing transmitter with the release increasing the more the horizontal cells are depolarized by the photoreceptors.

The basis for our deductions includes the following observations contributed by electrophysiological analysis. The photoreceptors release a depolarizing transmitter, and the release increases as the intensity of the light stimulus is reduced. The horizontal cells release a hyperpolarizing transmitter, and the release increases when the horizontal cells are depolarized by the photoreceptors following a reduction in the light intensity. We conceive of the bipolar cells as being exposed to a continuous depolarizing pressure exerted by the photoreceptors and a continuous hyperpolarizing pressure exerted by the horizontal cells. The balance between these two opposing influences determines the membrane potential of the bipolar cells at each instant. Also, we assume that there is a limited range within which the membrane potential can change.

Our theoretical deductions involve neural interactions at the synaptic connections of dendritic endings of bipolar cells, that is, at a structural level where no recording is possible. However, we can compare the deductions with the different types of responses that have been recorded from individual bipolar cells, responses that reveal the end result of the neural interaction. What the theoretical deductions may contribute is an understanding of the neural events that together have shaped the bipolar cell responses. It is these neural events that constitute information processing, while the response of the neuron reveals the end result of the processing at a particular level within a neural center.

If the deduced and the recorded responses agree, they support our approach to analyzing information processing in neural centers.

2. Summary of the Different Patterns of Synaptic Connections of Bipolar Cells

At cone terminals we can distinguish among four patterns of synaptic connections of bipolar cells. Two of these patterns are characterized by the bipolar cell contributing endings making triad-type connections at synaptic ribbon complexes. According to one of these two patterns these endings are the only endings of the bipolar cell at the terminal, while according to the second pattern the bipolar cell contributes large endings received in separate invaginations in addition to the small endings at synaptic ribbon complexes.

Functionally, the endings at synaptic ribbon complexes must be exposed to the influence of the photoreceptor as well as to that of horizontal cells. The endings that are separated from the synaptic ribbon complexes, on the other hand, are influenced by the photoreceptor only.

A third pattern of bipolar cell connections is characterized by the endings being located in invaginations outside the invaginations containing the synaptic ribbon complexes and the bipolar cells being connected to only one of the horizontal cells at synaptic ribbon complexes through communications between the invaginations at the sclerad part of the complexes. In these cases, the conditions are fulfilled for the photoreceptor dominating in influence on the bipolar cell, while the horizontal cells may modify the bipolar cell responses.

According to the fourth pattern the bipolar cell endings are received in invaginations completely separate from the invaginations accomodating the synaptic ribbon complexes, and the bipolar cells are not connected to any horizontal cells at the terminal. One such bipolar cell at terminal 1 was connected to two small horizontal cells just vitread to the terminal. Another bipolar cell made no connection to horizontal cells at all. This fourth pattern must favor the photoreceptor's determining the bipolar cell response, and the influence from horizontal cells is either absent or of a special nature, mediated by small horizontal cells.

At rod terminals we can distinguish between bipolar cell endings at synaptic ribbon complexes with extensive triad-type connections to the two horizontal cells and bipolar cell endings located in a separate invagination extending laterally from the synaptic ribbon complex. This second type of connection appears similar to the third pattern of bipolar cell connections at cone terminals. The first pattern of rod connections allows the horizontal cells to exert a strong influence on the transmission between the rod and the bipolar cell. In contrast, the second pattern favors the rod's predominantly determining the response, while horizontal cells are capable of modifying the response.

A third pattern is similar to the second pattern at cone terminals. In this case one end branch makes an asymmetric triad connection at the synaptic ribbon complex, whereas a second end branch with no horizontal cell processes contacting it connects to the vitread surface of the rod terminal. The first end branch is exposed to horizontal cell influence, whereas the second end branch is influenced by the rod exclusively.

According to a fourth pattern bipolar cell endings contact the surface of the rod terminal at its vitread pole without being invaginated. There may be several such connections at one rod terminal. This pattern is similar to the fourth pattern of bipolar cell connections at cone terminals.

3. One Design Principle: Saving Space

The fact that the endings of the bipolar cells and of the horizontal cells fit closely in deep invaginations of the photoreceptor membrane is one feature that must reflect an adaptation of the design to a particular functional requirement. One consequence of the invagination of the neural endings is efficient shielding of the neural connections, reducing the possibility that transmitter released by nerve endings outside the synaptic ribbon complex can interfere with the transmission at the complex. Usually this kind of shielding is established by glia cells in the central nervous system and by Müller's cells in the retina.

The shielding by invagination reduces the space required for the connections. We therefore conclude that at this level the circuitry is adapted to the saving of space. Rather complex circuitry can therefore be located at the level of the photoreceptors in spite of their close arrangement.

4. The Information Contributed by the Horizontal Cells

To evaluate the input to a bipolar cell from large horizontal cells, we must know the size, shape, and location of the area within which the large horizontal cells receive input from photoreceptors that determines their influence on that bipolar cell. The large horizontal cell processes extend long distances in one direction. They are connected to the photoreceptor terminals by short, thin branches. The area from which a large horizontal cell process receives input from photoreceptors is therefore a narrow zone extending along the process.

The input to bipolar cells from these processes depends on the length of the zone over which the photoreceptor input to the horizontal cell process affects the bipolar cell. This length is determined by the extent to which the potential change imposed by the photoreceptors becomes attenuated when conducted along the process. The attenuation has been determined in large horizontal cell processes that are accessible to such analysis. The amplitude of the potential change decreases exponentially with the distance from its site of initiation (Naka and Rushton, 1967). With such rapid attenuation it can be expected that the range over which a potential change is conducted without a major reduction of amplitude is limited.

These recordings were made on the very large cylindrical processes of the large horizontal cells. The attenuation must be even more pronounced in the much thinner processes of the small horizontal cells. The tapered shape of these processes must also contribute to the attentuation of a potential change conducted toward the soma of the horizontal cell.

Such rapid attenuation seems to preclude the membrane potential of an entire large horizontal cell from always assuming a uniform level. Instead, it seems likely that the potential can differ in different processes and even within different parts of a large process. The input to a bipolar cell is then determined by the input that the large horizontal cells receive from a certain limited area. This area we define as the functional receptive field of the horizontal cell relative to the bipolar cell. The input to the horizontal cell within its functional receptive field we consider to be averaged.

We conclude that the rapid attenuation of the potential change conducted along a large horizontal cell process restricts the functional receptive field to individual processes. This means that the functional receptive field of one large horizontal cell process is elongated and transmits information to a bipolar cell regarding the illuminance over a restricted area of the retina.

One such process therefore is insufficient to contribute information about the ambient illuminance. The combination of input from two large horizontal cell processes oriented perpendicularly, however, improves the conditions for the input reflecting the ambient illuminance.

This reasoning offers an explanation for the observation that bipolar cells connected to large horizontal cells are connected to processes approaching the terminal from two roughly perpendicular directions. The only exceptions from this rule involved bipolar cells with too few connections at the reconstructed terminals to allow evaluation of the overall pattern of horizontal cell connections.

We interpret the observations as revealing that one input to the bipolar cells from the horizontal cells contributes information regarding the ambient illuminance and that this information is basic to the shaping of the responses of the majority of the bipolar cells.

In contrast to the large horizontal cells, the information transmitted by the small horizontal cells concerns the illumination of a small area of the retina located in one particular direction from the photoreceptor–bipolar cell synapse. These horizontal cells therefore contribute information regarding a pattern of distribution of light over the receptive field of the bipolar cell.

5. Prevention of Mutual Interference of Transmission at Synaptic Ribbon Complexes in Cone Terminals

The differences in the neural connections of bipolar cells at the cone terminals allow the bipolar cells to transmit different types of information to the inner plexiform layer. This requires however, that there be no mutual interference at the synaptic connections of the bipolar cells at the terminals.

Such autonomy is guaranteed by the endings being located in deep invaginations that open at the basal surface of the terminal with narrow openings. Any effect exerted by horizontal cells on the photoreceptor at synaptic ribbon complexes will cause a potential change in the photoreceptor membrane that will not spread to other invaginations because the potential change attenuates rapidly when conducted over the basal surface of the terminal. We therefore conclude that the neural interactions at the large number of synaptic connections of bipolar cells at cone terminals occur simultaneously and independently according to different patterns determined by the combinations of neural connections.

6. Depolarizing Bipolar Cells

We distinguished between a vitread and a sclerad part of the synaptic ribbon complexes. In the vitread part the bipolar cells are connected to the photoreceptor and to two horizontal cells, while in the sclerad part the neural connections are confined to the photoreceptor and the two horizontal cells.

The following invariable structural features must play a role in transmission at the synaptic ribbon complexes. In the vitread part of the complex the horizontal cell endings are strikingly thin and abruptly widen as they extend into the sclerad part of the invagination. In cone terminals the bipolar cells are in contact only with the thin part of the horizontal cell endings, showing that the transmission between the horizontal cells and the bipolar cell does not require endings of a diameter larger than that of the thin part of the horizontal cell endings. We then want to find an explanation for the widening of the horizontal cell endings in the sclerad part.

In addition, we want to explain why the two horizontal cell endings are located in separate branches of the invagination. This separation allows the interaction between one horizontal cell and the photoreceptor to occur independently of the corresponding interaction of the other horizontal cell. This independence is enhanced further by the presence of the synaptic ribbon, which is situated so it can act as a shield between the two horizontal cell endings.

The presence of synaptic vesicles both in the horizontal cell endings and in the photoreceptor terminals shows that conditions for a reciprocal synaptic interaction between the photoreceptor and the horizontal cells are fulfilled.

The next structural feature that can be related to function is the presence of a zone 0.2 μm wide extending from the edge of the septum and within which there are no synaptic vesicles on the photoreceptor side of the connection. This is the only region of cytoplasm adjacent to the photoreceptor membrane in which there are no synaptic vesicles.

With the conditions for a reciprocal synapse fulfilled we expect to find an area of the photoreceptor membrane that is postsynaptic to the horizontal cells. The absence of synaptic vesicles at a part of the septum, we conclude, indicates the postsynaptic area of the photoreceptor membrane. The presence of cytoplasmic densities at the neural membranes facing each other within this zone is also an indication that this is a special region differing from the rest of the photoreceptor–horizontal cell connection.

If we now assume that the horizontal cells release their hyperpolarizing transmitter within the vitread part of the complex, both the bipolar cell and the photoreceptor will be exposed to a hyperpolarizing pressure (Figs. 3.1A and B). The hyperpolarization of the photoreceptor involves an area of the photoreceptor membrane located at the entrance to the two branches of the invagination that extend sclerad. A potential change imposed on the photoreceptor membrane at this location consequently is conducted sclerad along the invaginated membrane. The abrupt widening of the invagination must lead to a considerable attenuation of the conducted potential change (Fig. 3.1C).

The consequence is that the hyperpolarizing pressure exerted on the photoreceptor is greatest in the vitread part of the photoreceptor membrane and considerably lower in the sclerad part. The horizontal cells can then affect the transmission between the photoreceptor and the bipolar cell in two ways. In addition to the hyperpolarizing pressure exerted on the bipolar cell directly by the horizontal cells, the latter cells oppose the depolarizing action of the photoreceptor on the bipolar cell through the hyperpolarizing pressure they exert locally on the photoreceptor. This pressure is maximal at the synaptic connections between the photoreceptor and the bipolar cell and greatly reduced at the connection between the photoreceptor and the horizontal cells.

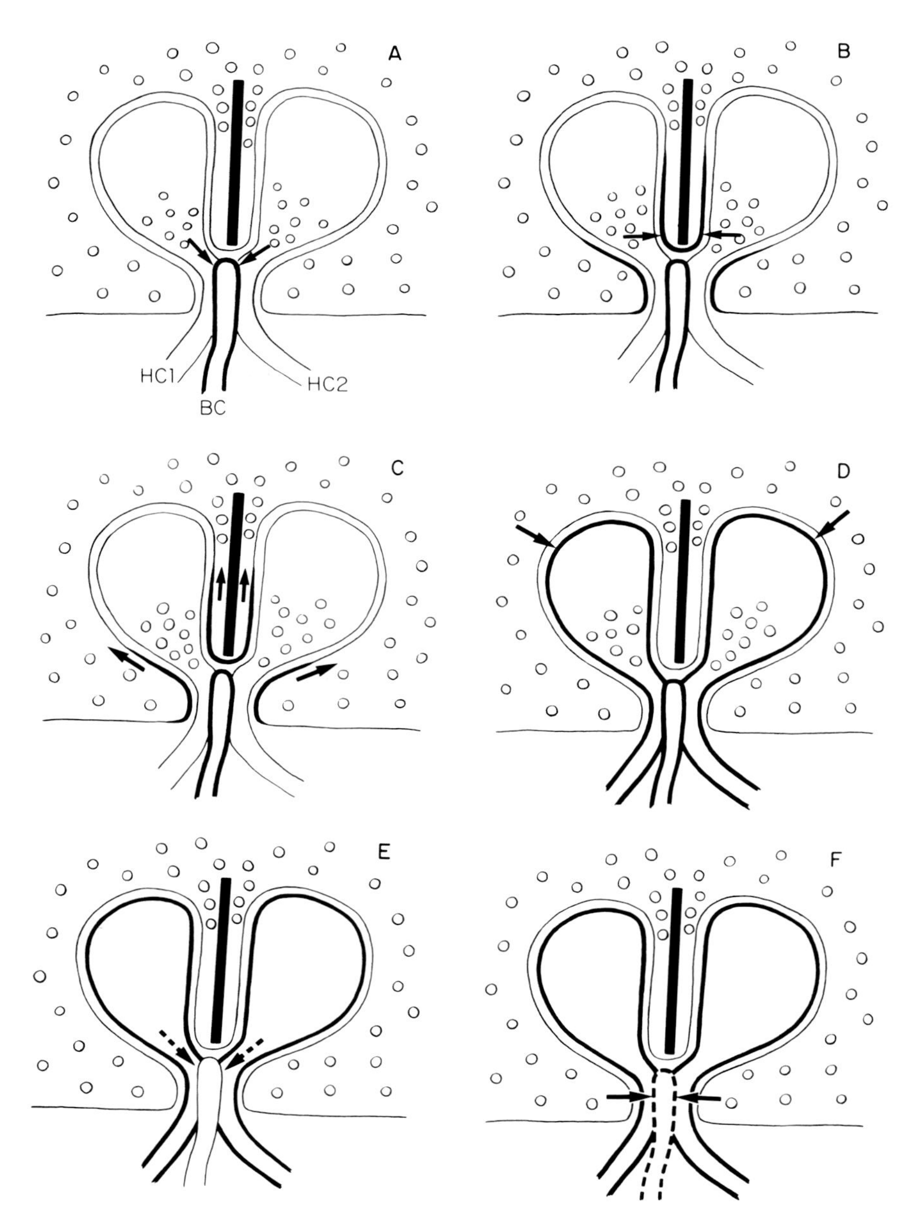

Fig. 3.1. Pictorial description of the synaptic mechanism that regulates the on-responses of a bipolar cell. Horizontal cells block the transmission between the photoreceptor and the bipolar cell by exerting a strong hyperpolarizing pressure on the bipolar cell (A) and on the photoreceptor (B). The latter effect is confined to the vitread part of the complex. (C) The

In the sclerad part of the invagination synaptic vesicles are present in the photoreceptor cytoplasm, and cytoplasmic densities may be located at the photoreceptor membrane. These are structural conditions for the photoreceptor being the presynaptic component.

On the basis of this structural organization of the synaptic ribbon complex we deduce the following function of the complex. The connections in the vitread part of the complex allow the horizontal cells to block the transmission between the photoreceptor and the bipolar cell by locally imposing a hyperpolarization block. The transmission of a signal then requires that the hyperpolarizing pressure of the horizontal cells be reduced, which occurs when the photoreceptor is stimulated by light. The light stimulus reduces the rate at which the photoreceptor releases the depolarizing transmitter, pushing the horizontal cell ending toward hyperpolarization (Fig. 3.1D). The consequent reduction in the hyperpolarizing pressure exerted by the horizontal cells on the photoreceptor and on the bipolar cell breaks the hyperpolarization block. The photoreceptor can then depolarize the bipolar cell because it still releases the depolarizing transmitter, although at a somewhat reduced rate (Fig. 3.1E). It is not the absolute rate at which this transmitter is released but the rate of release relative to the rate of release of the hyperpolarizing transmitter by the horizontal cells that determines whether there will be a hyperpolarization block of the transmission or whether the photoreceptor will depolarize the bipolar cell.

When the light stimulus ceases, the hyperpolarizing pressure is reestablished at a level where the transmission between the photoreceptor and the bipolar cell is blocked. Thus the horizontal cells function like a gate that opens as a light stimulus is being turned on and closes as it is shut off.

---

imposed hyperpolarization spreads along the two branches of the invagination while undergoing attenuation as a consequence of the abrupt and extensive widening of the invagination at the two branches. This attenuation prevents a blocking of the transmission between the photoreceptor and the horizontal cells. Instead, the photoreceptor exerts a continuous depolarizing pressure on the horizontal cells. The magnitude of this pressure is determined by the intensity of the light acting on the photoreceptor and by the overall input to the horizontal cells from photoreceptors. The latter input obviously contributes to the hyperpolarizing pressure exerted by the horizontal cells. When light stimulates the photoreceptor, it reduces its depolarizing pressure and the horizontal cell endings are driven toward hyperpolarization (D). If the consequent reduction in the hyperpolarizing pressure maintained by the horizontal cells is extensive enough (E), the hyperpolarization block is broken and the photoreceptor can depolarize the bipolar cell (F). The thick lines representing the neural membranes indicate that the membrane is driven toward hyperpolarization, and the broken lines indicate that the membrane is driven toward depolarization. Arrows with broken lines indicate a reduction in the exerted hyperpolarizing pressure.

Bipolar cells connected to synaptic ribbon complexes, we conclude, respond with depolarization when light stimulates the photoreceptors to which they are connected. These bipolar cells we refer to as depolarizing bipolar cells and to their response as on-response.

According to our deduction the shape of the invagination accommodating synaptic ribbon complexes is of crucial importance to the horizontal cell's function. This explains why when a horizontal cell ending participates at two synaptic ribbon complexes arranged in series, the ending consists of two widened parts separated by a narrow section. Both widened parts reflect the shape of the sclerad part of the invagination at the two complexes and the interposed narrow section reflects the narrow vitread part of the invagination at the sclerad complex. At both complexes the same abrupt change in the width of the process and consequently of the invagination is associated with the transition between the vitread and the sclerad parts of the complexes allowing the same attenuation of the horizontal cell's influence on the photoreceptor at both complexes.

7. Hyperpolarizing Bipolar Cells

Some bipolar cells both at cone and rod terminals did not send any endings to the synaptic ribbon complexes to make connections of the type described in the previous section. Of these bipolar cells some were connected to the horizontal cells outside the synaptic ribbon complexes, making lateral connections, while others made no connections to horizontal cells at all.

In these cases the bipolar cell becomes hyperpolarized when the photoreceptor is stimulated by light; the hyperpolarization is reversed when the light stimulus is shut off. These bipolar cells are therefore hyperpolarizing bipolar cells.

8. Bipolar Cells Depolarized When a Light Stimulus Is Both Turned On and Shut Off

Four bipolar cells connected to terminal 1 contributed endings to synaptic ribbon complexes as well as endings that were located in invaginations separate from the invaginations containing synaptic ribbon complexes. These bipolar cell endings therefore are structurally qualified to respond with depolarization both when the photoreceptor is stimulated by light and when the light is shut off.

In this manner we can distinguish between three types of responses of bipolar cells that can be deduced on the basis of the structural organization of the neural endings at the photoreceptor terminals. These three types apply to both cone and rod terminals. In rod terminals one bipolar

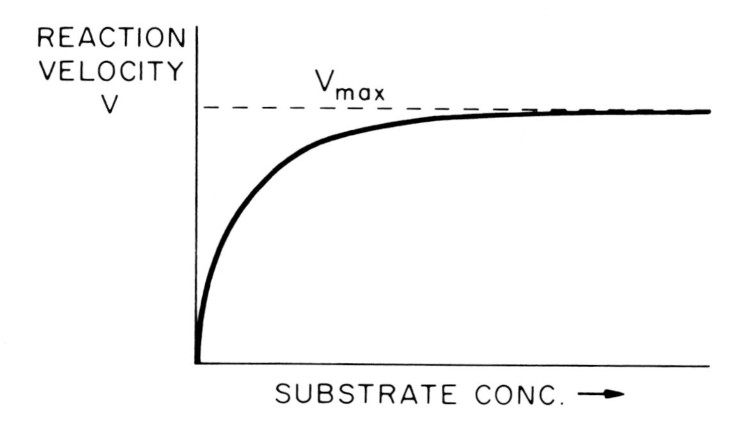

Fig. 3.2. Michaelis–Menten plot.

cell was always found to make an extensive triad type of connection at the synaptic ribbon complex, while the ending of the other bipolar cell made connections according to one of the other two patterns. Of these two patterns we interpret the lateral connections to horizontal cells as representing the pattern of connections of a hyperpolarizing bipolar cell. Bipolar cells that in addition to contacting the synaptic ribbon complex also contact the vitread pole of the rod terminal with a separate ending we consider to be depolarized both when the light stimulus is turned on and when it is shut off.

9. Limitations to the Horizontal Cells' Influence on Bipolar Cells

Let us now assume that the interaction of the transmitter and the receptor protein can be compared to a substrate–enzyme interaction for which the Michaelis–Menten equation applies. A plot of this equation (Fig. 3.2) shows that at constant enzyme concentration the reaction rate increases with increasing substrate concentration according to a curve that asymptotically approaches a maximum. This maximum corresponds to a concentration that keeps the enzyme saturated with substrate. The plot shows that at high substrate concentrations there is a wide range within which large changes in the concentration cause only small changes in the reaction velocity.

According to this analogy the binding of the transmitter to the receptor protein imposes a change in ion permeability corresponding to the change of the substrate molecule imposed by its binding to an enzyme molecule.

The Michaelis–Menten relationship should apply because there is a fixed concentration of receptor protein in the postsynaptic membrane.

The rate at which the transmitter is bound to the population of receptors increases as the release of transmitter is increased until the receptors are saturated with transmitter. When this stage is being approached, a progressively larger increase in the release of the transmitter is required to cause a given change in the interaction between receptor protein and transmitter and consequently to cause a given change in the membrane potential of the bipolar cell.

The increments in the stimulus intensity must consequently be larger and larger to evoke a potential change of a given amplitude across the bipolar cell membrane when the horizontal cells approach maximum depolarization at a low illuminance. We can then expect that within a rather wide range, as shown by Fig. 3.2, the hyperpolarizing pressure exerted by the horizontal cells remains practically constant, and we can distinguish between two ranges of illuminance. Within one range the horizontal cells modulate the hyperpolarizing pressure with the amplitude of the variation being large enough to evoke corresponding changes in the polarization of the bipolar cell and of the photoreceptor membranes. Within a second, lower range of illuminance the variations in the release of hyperpolarizing transmitter do not evoke any appreciable variations in the polarization of the bipolar cell and of the photoreceptor membranes.

With the hyperpolarizing pressure exerted by the horizontal cells on the bipolar cells balancing the depolarizing pressure exerted by the photoreceptors, we conclude that the bipolar cells respond in different ways within these two ranges of illuminance. At a low illuminance, the horizontal cells will exert a constant, non-modulated, hyperpolarizing pressure on the bipolar cells' membrane potential.

10. The Receptive Field of the Bipolar Cell

A bipolar cell receives information directly from photoreceptors to which it is synaptically connected, and in addition it receives information indirectly through the connections to horizontal cells. The area over which the horizontal cells transmit information to a bipolar cell is the receptive field of the bipolar cell. Within this receptive field we can distinguish a center, within which the bipolar cell in addition is influenced directly by photoreceptors, and a surround, from which information from photoreceptors is mediated by horizontal cells only.

We can then deduce that the receptive field of a bipolar cell varies in size with the illuminance because the functional receptive fields of the horizontal cells vary as a consequence of the attenuation of the potential changes conducted along the horizontal cell processes and the distance over which a potential change will be conducted before the amplitude is greatly reduced depends on the original amplitude of the potential change and consequently is determined by the illuminance.

11. The Balance between Photoreceptor and Horizontal Cell Influence on Bipolar Cells, the Baseline Potential

We have pointed out that the potential difference across the bipolar cell membrane is determined by two opposing influences, the depolarizing pressure exerted by the photoreceptors and the hyperpolarizing pressure exerted by the horizontal cells. Under steady-state conditions these pressures make the bipolar cell membrane assume a potential somewhere between maximal depolarization and maximal hyperpolarization. This steady-state potential we refer to as the baseline potential.

This potential is not a fixed potential such as a resting potential encountered under firmly controlled artifical experimental conditions. The reason why we do not expect the baseline potential to be fixed is the fact that there is no simple coupling of the hyperpolarizing pressure to the depolarizing pressure in the steady state. Depending on the distribution of light over the receptive field of a bipolar cell, the hyperpolarizing pressure exerted by the horizontal cells can vary to a certain extent independently of the depolarizing pressure associated with the light distribution over the center of the receptive field. As a consequence the baseline potential must vary; it will be a floating potential.

12. The Type 1 and Type 2 Responses of a Bipolar Cell

Let us now deduce in more detail the responses of the bipolar cells to variations in the illuminance. First, consider a hyperpolarizing bipolar cell with connections to horizontal cells and the situation at a low ambient illuminance, when the horizontal cells' modulating influence is minimal. These responses can then be predicted to be simple. When the center of the receptive field of a hyperpolarizing bipolar cell is illuminated, the cell becomes hyperpolarized by the photoreceptors, and the change in membrane potential is sustained during the light stimulus. When the light is shut off, the membrane potential returns to its original level. This is the type 1 response of a bipolar cell (Fig. 3.3).

If we gradually increase the intensity of the ambient illuminance, we eventually reach intensities at which the hyperpolarizing pressure changes appreciably when the light intensity varies. The horizontal cells can now modulate their influence on the photoreceptors and the bipolar cell. A light stimulus makes the photoreceptors reduce their depolarizing

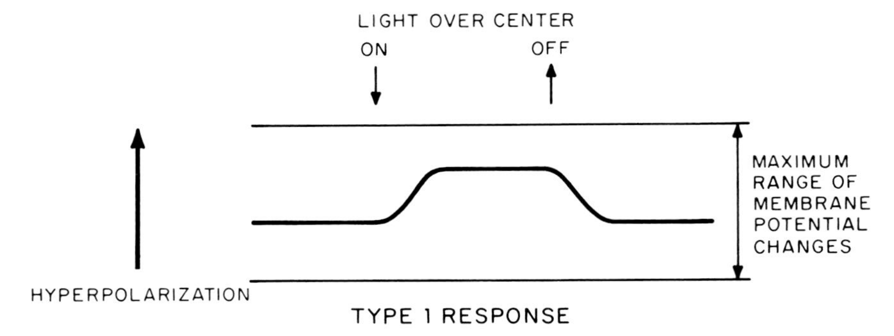

Fig. 3.3. The type 1 response of a bipolar cell. The cell is hyperpolarized by the photoreceptor without horizontal cells balancing the response.

pressure, while during a synaptic delay the hyperpolarizing pressure exerted by the horizontal cells remains unchanged. The bipolar cell membrane potential during this delay is therefore pushed toward hyperpolarization by the horizontal cells against a reduced depolarizing pressure exerted by the photoreceptors. After a synaptic delay, however, the horizontal cells reduce their hyperpolarizing pressure as a consequence of the photoreceptor influence. The depolarizing pressure exerted by the photoreceptors then pushes the bipolar cell membrane potential toward depolarization, partially reversing the initial hyperpolarization. If the light stimulus is maintained at the same intensity, the bipolar cell membrane potential remains at a baseline level somewhere above the level of the baseline potential before the light was turned on. The extent to which the bipolar cell membrane potential is reversed depends on the magnitude of the change in the hyperpolarizing pressure exerted by the horizontal cells.

When the light is shut off, the photoreceptors increase their depolarizing pressure and the bipolar cell membrane potential is pushed toward depolarization against the at first unchanged hyperpolarizing pressure exerted by the horizontal cells. After a synaptic delay this pressure is restored to its higher level before the light was turned on, and the consequent increase in pressure pushes the bipolar cell membrane toward hyperpolarization and back to the original baseline potential, partially reversing the depolarization of the bipolar cell. The response of the bipolar cell is therefore depolarization, transiently overshooting the baseline potential. This is the type 2 response of a bipolar cell (Fig. 3.4). Interference by the horizontal cell thus changes a sustained hyperpolarization to a transient hyperpolarization followed by a transient deflection toward depolarization. The fact that this deflection overshoots the original baseline potential shows that the modulating influence of the

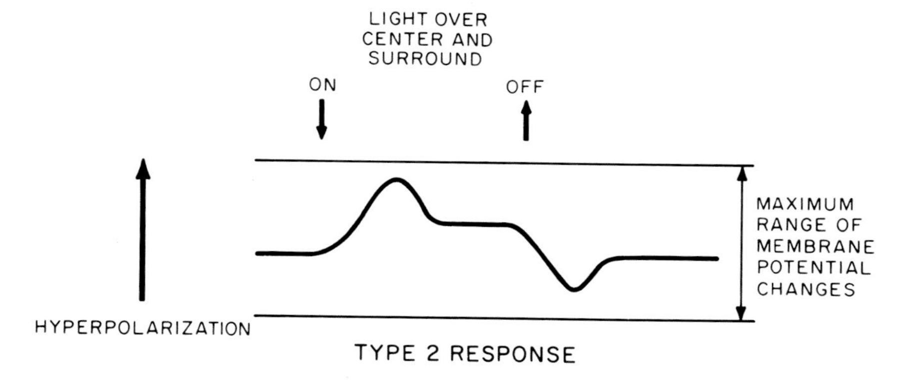

Fig. 3.4. The type 2 response of a bipolar cell. Horizontal cells interfere in the response.

horizontal cell increases the amplitude of the potential change toward depolarization.

This horizontal cell's interference, however, applies only to conditions under which the horizontal cells can vary the hyperpolarizing pressure. It does therefore not apply to a low ambient illuminance except for high-intensity stimuli, and it does not apply when the horizontal cells approach maximal hyperpolarization at a high ambient illuminance. In the latter case the extensive hyperpolarization of the horizontal cells decreases the range over which the horizontal cells can reduce the release of hyperpolarizing transmitter when the photoreceptors are stimulated by light at intensities above the ambient illuminance.

13. The Center and the Surround of the Bipolar Cell Receptive Field: The Type 3 Response

By confining the changes in the light stimuli to the surround we can deduce the responses of a bipolar cell evoked by the horizontal cells only. When the intensity of the light over the surround is increased while it is kept constant over the center, the hyperpolarizing pressure on a bipolar cell decreases as a consequence of the hyperpolarization of the horizontal cells. With the depolarizing pressure exerted by the photoreceptors directly on the bipolar cell unchanged, the membrane potential of a bipolar cell is pushed toward depolarization. A hyperpolarizing bipolar cell can therefore be depolarized either by the light intensity over the center being reduced or by the light intensity over the surround being increased.

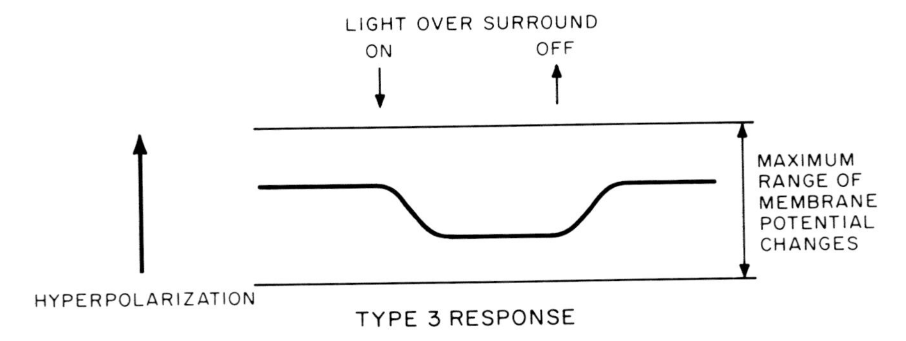

Fig. 3.5. The type 3 response of a bipolar cell. It is generated by varying the illumination over the surround.

In the opposite case, when the intensity of the light is decreased over the surround but kept constant over the center, the hyperpolarizing pressure on the bipolar cell increases as a consequence of the depolarization of the horizontal cells. The bipolar cell membrane potential is therefore pushed toward hyperpolarization. We refer to these responses that illustrate the influence exerted by the horizontal cells without interference of changes in the intensity of the light stimulating the photoreceptors in the center as the type 3 responses (Fig. 3.5).

14. Sign-Reversing Neurons

When a horizontal cell is depolarized by photoreceptors it pushes the membrane potential of bipolar cells to which it is connected toward hyperpolarization. This means that the original depolarizing influence of the photoreceptors has been changed to a hyperpolarizing influence. The sign of the original potential change has been reversed. This reversal of the sign is accomplished by a neuron interposed between the photoreceptor and the bipolar cell releasing a hyperpolarizing transmitter.

Obviously, a horizontal cell can only increase or decrease the hyperpolarizing pressure exerted on a bipolar cell; it cannot induce a reversal of hyperpolarization to depolarization. Such a sign reversal requires the depolarizing influence exerted directly by photoreceptors on bipolar cells, as in the case of the depolarizing bipolar cells.

In most cases both large and small horizontal cells are connected to each bipolar cell. Small and large horizontal cells are connected through special end branches extending from either the small or the large cell, and synaptic vesicles are accumulated in the large horizontal cell processes at the site of the contact. Thus the large horizontal cells are connected synaptically to the small horizontal cells, and they are the presynaptic components.

If we now include the small horizontal cells in our scheme, we find that these cells can act as sign-conserving neurons in a chain of three neurons consisting of the photoreceptors, a large horizontal cell, and a small horizontal cell. For instance, if the surround is illuminated, the large horizontal cells become hyperpolarized and therefore reduce their hyperpolarizing pressure on the small horizontal cells. The reduction of this pressure allows the depolarizing pressure exerted by the photoreceptors connected to the small horizontal cells to push these cells toward depolarization. As a consequence, the small horizontal cells increase their release of hyperpolarizing transmitter at their connections to bipolar cells. The hyperpolarizing influence of the photoreceptors on the large horizontal cells is therefore conserved at the small horizontal cells' connections to bipolar cells.

The conserved hyperpolarizing influence on the bipolar cells, however, is delayed by one synaptic delay during which the large horizontal cells expose the bipolar cells to a reduced hyperpolarizing pressure.

15. Bipolar Cells Connected to Both Large and Small Horizontal Cells: The Type 4 Response

Our deduction of the effects of the horizontal cells on the membrane potential of the hyperpolarizing bipolar cell was based on the assumption that all horizontal cells exert the same influence on the bipolar cell. This applies to situations when the hyperpolarizing bipolar cell is connected only to large horizontal cells. When small horizontal cells are involved also, they contribute to a different modulation of the bipolar cell responses as a consequence of their connections to large horizontal cells.

Let us consider a hyperpolarizing bipolar cell that is connected to both large and small horizontal cells when the intensity of the light falling on the surround increases but remains constant over the center. The bipolar cell will be driven toward depolarization because the hyperpolarizing pressure maintained by both types of horizontal cells on the bipolar cell and on the photoreceptors is reduced, while the depolarizing pressure exerted by the photoreceptors is maintained constant.

However, after a synaptic delay, the reduction in the hyperpolarizing pressure exerted by the large horizontal cells on the small horizontal cells makes the latter cells increase their release of hyperpolarizing transmitter. As a consequence, the depolarization of the bipolar cell is partially reversed (Fig. 3.6).

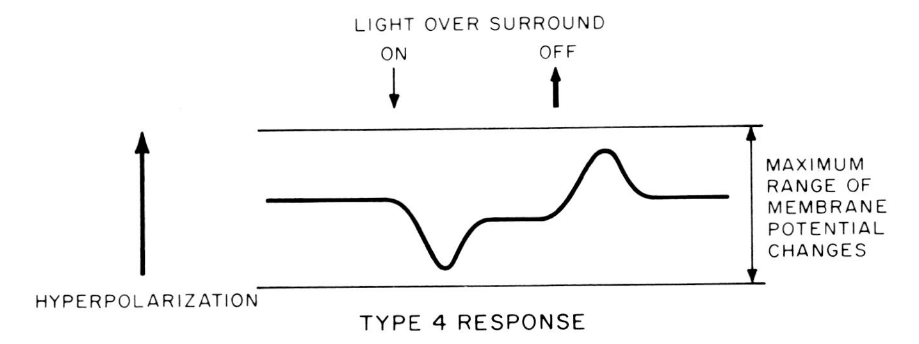

Fig. 3.6. The type 4 response of a bipolar cell. This response is shaped by the combined influence exerted by large and small horizontal cells when the intensity over the surround is varied.

The opposite applies to the situation when the intensity of the light over the surround is decreased. The bipolar cell is then driven toward hyperpolarization by the increased hyperpolarizing pressure exerted by the horizontal cells until, after a synaptic delay, the consequent decrease in the release of hyperpolarizing transmitter by the small horizontal cells partially reverses the hyperpolarization (Fig. 3.6). These responses of the bipolar cell will be referred to as the type 4 responses. They represent a modification of the type 3 responses that applies to hyperpolarizing bipolar cells connected only to large horizontal cells.

16. Depolarizing Bipolar Cells: The Type 5, 6, and 7 Responses

We have deduced that the structure of the synaptic ribbon complexes allows the horizontal cells to control the transmission between the photoreceptor and the bipolar cell ending. As a consequence, the bipolar cell is depolarized when the photoreceptor is stimulated by light. When the light stimulus is turned off, the depolarization is reversed. This is the type 5 response of a bipolar cell (Fig. 3.7).

When horizontal cells become more involved in the response of the depolarizing bipolar cell, as when light stimulates both the center and the surround, the responses of a depolarizing bipolar cell are more complex if the cell is connected to both large and small horizontal cells.

When only the surround is illuminated, the horizontal cells reduce their hyperpolarizing pressure at the synaptic ribbon complexes, and the bipolar cell is depolarized by the photoreceptors. This effect of illumination of the surround is identical to that of the center, and light over both

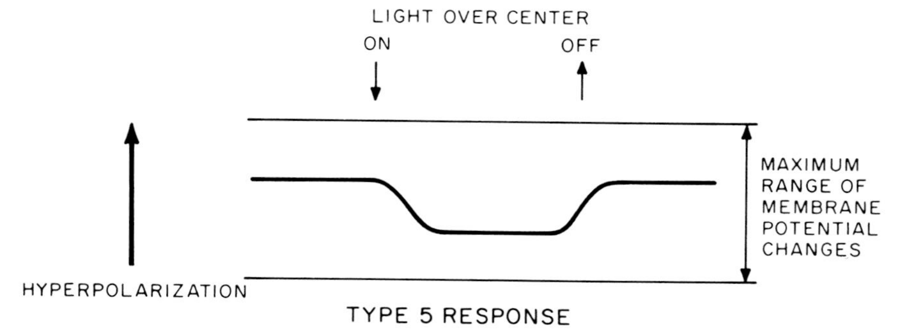

Fig. 3.7. The type 5 response of a bipolar cell. The depolarizing bipolar cell responds to changes in the illumination over the center. Horizontal cells interposed between the photoreceptor and the bipolar cell at the synaptic ribbon complexes function as sign-reversing neurons.

center and surround therefore enhances the depolarization of the bipolar cell.

However, after a synaptic delay, the small horizontal cells react to the reduction of the hyperpolarizing pressure exerted on them by the large horizontal cells, and as a consequence they increase their hyperpolarizing pressure at the synaptic ribbon complexes, reducing the depolarization of the bipolar cell. The bipolar cell response therefore is transformed into a transient depolarization.

When the light over the surround decreases, the hyperpolarization block of the transmission between photoreceptor and a depolarizing bipolar cell is reinforced until, after a synaptic delay, the increased hyperpolarizing pressure on the small horizontal cells reduces their contribution to the block. The bipolar cell membrane potential should therefore be deflected toward hyperpolarization when the light over the surround is turned off, and after a synaptic delay the potential change should be partially reversed. This is the type 6 response of a bipolar cell (Fig. 3.8).

When both the center and surround are illuminated, the surround first enhances the depolarization of the bipolar cell until the depolarization is partially reversed by the small horizontal cells. When the light stimulus is turned off, the potential across the bipolar cell membrane is pushed toward hyperpolarization by the horizontal cells while the depolarizing pressure exerted by the photoreceptors increases and more or less suppresses the hyperpolarization. The deflection of the membrane potential toward hyperpolarization should therefore be small or absent. This is the type 7 response of a bipolar cell (Fig. 3.9).

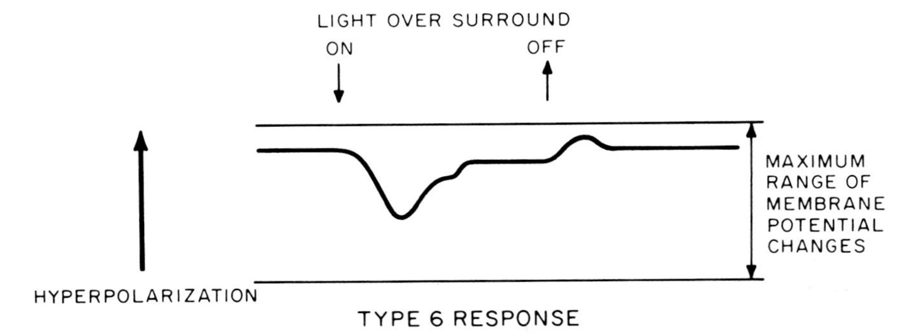

Fig. 3.8. The type 6 response of a bipolar cell. A depolarizing bipolar cell responds to changes in the illumination over the surround. The bipolar cell is connected to mixed synaptic ribbon complexes.

17. Why Are There Two Branches of the Invagination at Synaptic Ribbon Complexes?

The location of the bulbous sclerad part of the two horizontal cell endings in separate branches of the invagination of the photoreceptor membrane constitutes a most characteristic feature of the synaptic ribbon complex. In section 6 we deduced a functional significance for the abrupt widening of the horizontal cell endings and for the separation of one vitread part of the complex from a sclerad part. However, we have not yet deduced any functional significance for the separation of the two horizontal cell endings within the sclerad part. The structure clearly reveals that

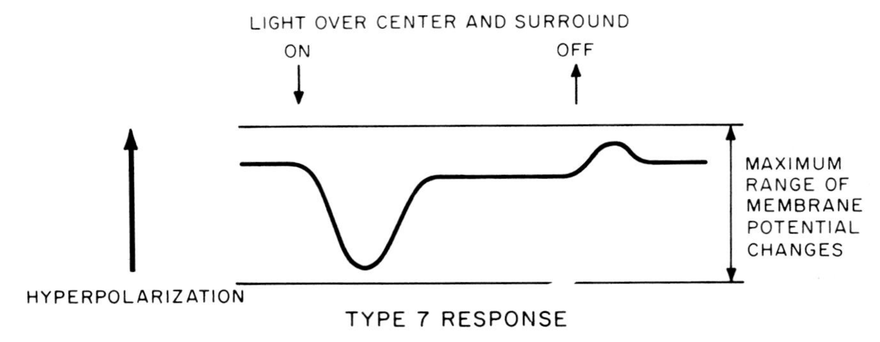

Fig. 3.9. The type 7 response of a bipolar cell. The depolarizing bipolar cell is connected to mixed synaptic ribbon complexes and responds to simultaneous changes in illumination over the center and the surround.

the two endings are separated preventing events at one ending from interfering with events at the other ending. The synaptic ribbon appears to form an efficient shield extending through the entire septum separating the two endings in cone synaptic ribbon complexes.

In the case of mixed synaptic ribbon complexes the separation of the horizontal cell endings allows the large and small horizontal cells to influence the transmission of signals from the photoreceptor to the bipolar cells independently.

There is one condition where such an independence appears to be of basic importance to the transmission between the photoreceptors and the bipolar cell. With the modulating influence of the horizontal cells on the bipolar cell responses vanishing at a low ambient illuminance, the extensively depolarized large horizontal cells maintain a hyperpolarizing pressure at a high and more or less constant level. The transmission between the photoreceptor and depolarizing bipolar cells could then be blocked.

However, because of the interaction between the two types of horizontal cells, the small horizontal cells will not be maximally depolarized, and we may therefore conclude that these horizontal cells can then modulate their hyperpolarizing pressure in a way required for the transmission between the photoreceptor and the bipolar cell. For this to occur the small horizontal cells must exert their influence on the bipolar cell and on the photoreceptor without the large horizontal cells interfering, because if the large cells interfere, the hyperpolarizing pressure exerted by the two horizontal cells would be averaged, and the pressure might well be high enough to sustain a hyperpolarization block preventing any transmission between the photoreceptor and the bipolar cell.

The branching of the invagination can prevent this averaging and thus permit the two horizontal cells to act independently on the two sides of the septum. One half of the septum can be kept extensively hyperpolarized while at the other half the photoreceptor can modulate the hyperpolarizing pressure exerted by the small horizontal cell sufficiently for transmission of signals to the bipolar cell even at low illuminance. This requires that the ion flux be maintained at different levels on the two sides of the septum. The synaptic ribbon, with its anchoring to the edge of the septum by a cytoplasmic density, seems to be in the position to furnish the diffusion barrier required to allow a temporary difference in the ion fluxes on the two sides of the septum. It is a proteinaceous structure dense enough to function as a diffusion barrier for ions. Therefore at low illuminance transmission of signals is possible only at mixed synaptic ribbon complexes and transmission to bipolar cells connected only to large horizontal cells at synaptic ribbon complexes is blocked.

The mutual contacts of the horizontal cells are another feature that characterizes the synaptic ribbon complex. In the cone terminal this contact is confined to the thin part of the horizontal cell endings in the vitread part of the complex. In contrast, there is no mutual contact between the corresponding parts in the synaptic ribbon complex in rod terminals. Instead, the widened part of the two horizontal cell endings in the sclerad part of the complex are in mutual contact within a zone vitread of the septum. The fact that the mutual contacts between the two horizontal cell endings are established according to two different patterns in rod and cone terminals gives strong support to the interpretation that these contacts are functionally important connections that may allow synaptic transmission.

When considering that there are synaptic connections of large and small horizontal cells in the outer plexiform layer, it appears reasonable to propose that corresponding synaptic connections are also established in the synaptic ribbon complex. The modulation of the state of polarization of the small horizontal cells by the large horizontal cells would then occur locally at the synaptic ribbon complex in addition to the modulation that occurs through the connections in the outer plexiform layer. This situation is comparable to the modulation of the bipolar cell potential by the horizontal cells through an input to horizontal cells from photoreceptors in the surround as well as through the local input at the synaptic ribbon complexes in the center.

Synaptic ribbons have been observed in the hair cells of the cochlea (Smith and Sjöstrand, 1961) and in bipolar cells in the inner plexiform layer (Kidd, 1962; Allen, 1969). In both of these cases the synapses appear to involve synaptic connections with synaptic vesicles located on both sides of the synaptic membranes, a situation similar to that of the reciprocal synapse at the synaptic ribbon complex. Also, the ribbon is oriented more or less perpendicular to the postsynaptic part of the neural membrane, and is frequently located opposite the site where the members of a pair of contacting endings are in mutual contact. The ribbon may also fulfill the function of a diffusion barrier at these synapses.

Thus we have related all the basic and constant structural features of the synaptic ribbon complexes to particular functional aspects of the transmission of signals between the photoreceptor and the bipolar cell ending. The complexity of the synaptic connections at synaptic ribbon complexes may be surprising. It shows that functionally important structural neural features are of dimensions far below the resolution of the light microscope. They can be revealed only by means of an electron microscopical analysis designed to disclose the three-dimensional arrangement of the neurons. We must therefore adapt to an entirely new level of dimensions in order to appreciate basic aspects of neural transmission associated with information processing.

18. Absence of Antagonism between Center and Surround: The Type 8 Response

We deduced that when a depolarizing bipolar cell is connected to mixed synaptic ribbon complexes, the surround contributes to the bipolar cell response positively at first, but that after synaptic delays the surround opposes the potential change imposed by the center. This opposing effect was instigated by the small horizontal cells, and it was the consequence of the interaction between the two types of horizontal cells. Depolarizing bipolar cells that are connected only to large horizontal cells should respond without the surround exerting any opposing influence on the depolarization of the bipolar cell membrane. Light over either the center or the surround should depolarize the bipolar cell. The depolarization should be maintained as long as the light stimulus lasts. When the stimulus is turned off the membrane potential should return to its original level. This is the type 8 response of a bipolar cell (Fig. 3.10).

The amplitude of a type 8 response thus reflects the average illuminance over the entire receptive field of the bipolar cell. Therefore, the spatial definition of the light stimulus will be poor. These bipolar cells, however, contribute to vision by continuously transmitting information regarding the average illuminance over the receptive field. At a low ambient illumination we expect that these bipolar cells will not function because of the strong hyperpolarization block maintained by both large horizontal cells at the synaptic ribbon complexes.

We will refer to these depolarizing bipolar cells as luminosity bipolar cells or L-bipolar cells.

Characteristically, the bipolar cells connected to several mixed synaptic ribbon complexes at terminal 1 also were connected to one synaptic ribbon complex at which both horizontal cell endings were contributed by large horizontal cells. These depolarizing bipolar cells should combine the transient depolarization when a light stimulus is turned on with a sustained, less extensive depolarization during the duration of the stimulus.

19. Bipolar Cells Depolarized When Light Is Both Turned on and Shut off: The Type 9 Response

The bipolar cells that are depolarized when light is both turned on and shut off are connected to the photoreceptor in two ways, by endings at

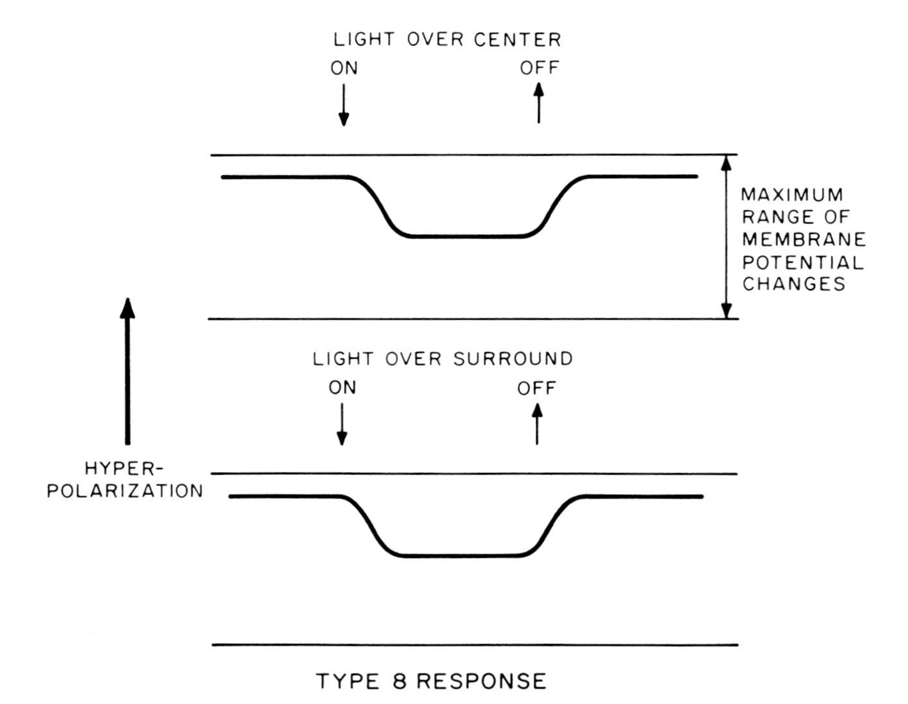

Fig. 3.10. The type 8 response of a bipolar cell. The depolarizing bipolar cell is connected only to synaptic ribbon complexes where both horizontal cells are large. The responses are the same whether the illumination is changed over the center, over the surround, or over both the center and the surround.

synaptic ribbon complexes and by endings outside these complexes. The latter endings are not in contact with any horizontal cells.

The connections at synaptic ribbon complexes make these bipolar cells respond with a transient depolarization when the photoreceptors are stimulated by light. The connections outside these complexes make them respond with depolarization when the light stimulus is shut off. The latter response is transient because the increased hyperpolarizing pressure exerted by the horizontal cells at the synaptic ribbon complexes partially reverses the depolarization of the entire ending. This effect is delayed by a synaptic delay. The responses of these bipolar cells will therefore be a combination of the type 6 and the type 2 responses, which we designate the type 9 response (Fig. 3.11).

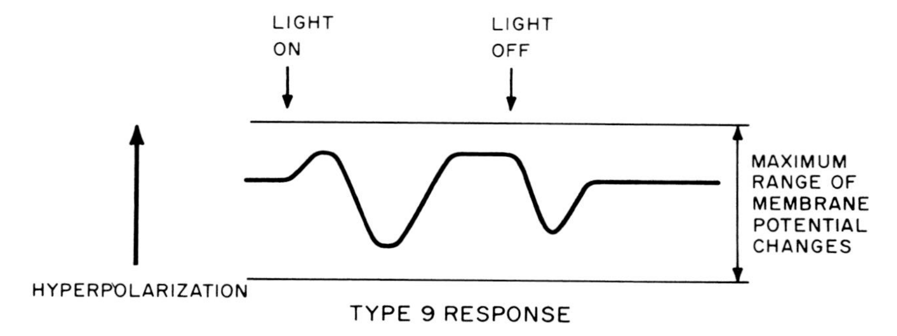

Fig. 3.11. The type 9 response of a bipolar cell: The on/off response.

20. Responses of Bipolar Cells Not Connected to Horizontal Cells: The Type 10 Response

The bipolar cells lacking connections to horizontal cells respond to a light stimulus with sustained hyperpolarization provided that their response is not modified by connections to other bipolar cells. Their state of polarization will be influenced little or not at all by changes in the illumination of the surround. When the light intensity over the center is reduced, the response will be depolarization that is sustained as long as the light intensity remains reduced. This is the type 10 response (Fig. 3.12).

The membrane potential of these bipolar cells floats with the input from the photoreceptors and therefore they can transmit information regarding the average level of illuminance over their receptive fields. If these fields are large, the transmitted information may reflect the ambient illuminance fairly accurately. When the receptive fields are small these bipolar cells transmit information regarding the illuminance locally.

21. The Different Functions of Bipolar Cells

We have deduced a considerable number of different responses of bipolar cells on the basis of the cells' synaptic connections at the photoreceptor terminals, and we have distinguished between depolarizing bipolar cells, hyperpolarizing bipolar cells, and bipolar cells responding to both increases and decreases in light intensity.

Typically, these responses transmit information regarding changes in the illuminance over the receptive fields of the bipolar cells. The

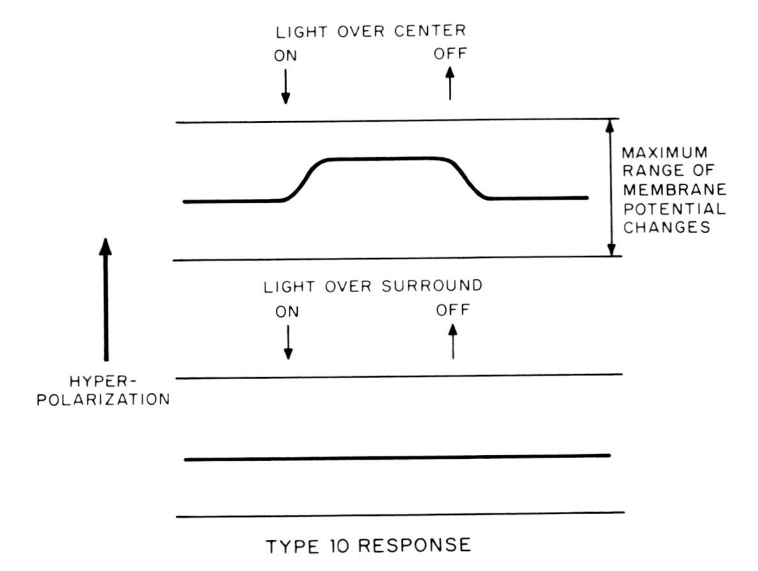

Fig. 3.12. The type 10 response of a bipolar cell. Because the bipolar cell is not connected to horizontal cells at the terminal, the responses are shaped entirely by the photoreceptor. Changes in the illumination over the surround have no influence on the response.

depolarizing bipolar cells in addition transmit information regarding the illuminance after the change has occurred, but the dominating response with the larger amplitude is the response to an increase in light intensity.

The L-bipolar cells, on the other hand, contribute information regarding the average illuminance over the entire receptive field, giving poor spatial definition of the stimulus.

All these bipolar cells are connected to large horizontal cells, and the ambient illuminance therefore plays a role in the modulation of the responses. This seems to qualify these bipolar cells for transmitting information on the basis of which a perceived image can be constructed in the brain. We can therefore conceive of these bipolar cells as being part of the channels that connect the photoreceptors to the visual centers in the brain.

The response to photoreceptor input by bipolar cells not connected to large horizontal cells is not modulated by the ambient illuminance. The photoreceptors exert on these bipolar cells a continuous depolarizing pressure that changes with the intensity of the light stimulating the photoreceptors. The pressure is high at low illuminance and gradually decreases as the illuminance increases. If these bipolar cells were transmitting information to image-transmitting ganglion cells they would cause a constant background noise leading to a corresponding deteriorating effect on the perceived image. Therefore these bipolar cells instead seem more suitable for exerting an influence on the transmission of signals within the retina by modulating neural interactions in the inner plexiform layer, where a control on the basis of the absolute magnitude of the input from photoreceptors may fulfill an important function.

The bipolar cells not connected to horizontal cells are characterized by their extensive connections to other bipolar cells. They therefore increase the complexity of the neural circuits. These bipolar cells may be justifiably distinguished from the other four types of bipolar cells as components of local circuits within the retina contributing to the integration of information within the retina. We will therefore refer to these bipolar cells as intrinsic bipolar cells. In this category of bipolar cells we also include the efferent bipolar cells, because they apparently act locally within the retina, and the core bipolar cells, on the basis of their extensive synaptic connections to other bipolar cells including connections to other intrinsic bipolar cells.

22. Absence of Inhibition of Bipolar Cells by Horizontal Cells

Even if the influence exerted on the bipolar cells by horizontal cells is of opposite sign to the influence exerted by the photoreceptors, the horizontal cells do not inhibit the bipolar cells. On the contrary, they contribute in a positive way by shaping the bipolar cell responses. For instance, consider depolarizing bipolar cells. Light falling on the surround of the receptive field will evoke depolarization, as will light falling on the center. The effect of the horizontal cells in this case is therefore of a facilitatory nature. The blocking of the transmission at synaptic ribbon complexes contributes to a reversal of the sign of the photoreceptor change to a light stimulus, and this horizontal cell action can not be considered inhibitory since it promotes the transmission of a signal to the bipolar cell. After a synaptic delay the small horizontal cells have been mobilized to increase their hyperpolarizing pressure, and these cells now exert an antagonistic influence that reduces the depolarization of the bipolar cell, but the antagonistic action of horizontal cells always follows a synaptic delay and contributes in a positive way by shaping the bipolar cell response. This modulating influence is also obvious in the case of hyperpolarizing bipolar cells, where it leads to an increase in the amplitude of the off response, a facilitatory influence.

The horizontal cells therefore do not qualify as inhibitory neurons, and a distinction between inhibitory and modulating neurons seems justifiable. Consequently, the horizontal cells cannot account for the observed inhibitory effects generated by light stimuli confined to areas away from the center of the receptive field of the bipolar cells. Such inhibition must be mediated by another type of neuron in the retina.

23. Oscillatory Responses

Because a depolarizing pressure is opposed by a hyperpolarizing pressure, and because the adjustment of one pressure to changes in the opposite pressure is delayed, a basis for the appearance of oscillations in the membrane potential of photoreceptors and horizontal cells seems to be created.

When the photoreceptors are stimulated by light, the depolarizing pressure is reduced, and as a consequence after a synaptic delay the hyperpolarizing pressure is reduced. This affects both photoreceptors and the bipolar cells, and the photoreceptors become more depolarized locally. The depolarizing pressure on the bipolar cells and horizontal cells increases, and the horizontal cells respond by increasing their hyperpolarizing pressure, pushing back the membrane potential of the photoreceptors locally toward hyperpolarization, and so on. It is the separation in time of the changes in the release of transmitters that seems to favor an oscillation of the membrane potentials of the photoreceptors, the horizontal cells, and the bipolar cells.

24. Comparison of the Deduced and the Observed Response Patterns of Bipolar Cells

The deductions made in the previous sections illustrate the aim of an analysis of the circuitry of neural centers based on a three-dimensional reconstruction of neural connections. Even the limited information considered so far has enabled us to propose functions for certain aspects of the circuitry of the retina. The extensive electrophysiological analysis of the retina allows us to compare these deductions with the recorded responses of bipolar cells and establish whether they agree with the observed responses.

Hartline (1938, 1940), studying small bundles of nerve fibers in the optic nerve, showed that the responses to light vary in different fibers. One type of response was a high-frequency discharge when the light was turned on followed by a lower-frequency response maintained until the light was turned off. A second type of response involved a burst of spikes when the light was turned both on and off. A discharge that appeared as the light

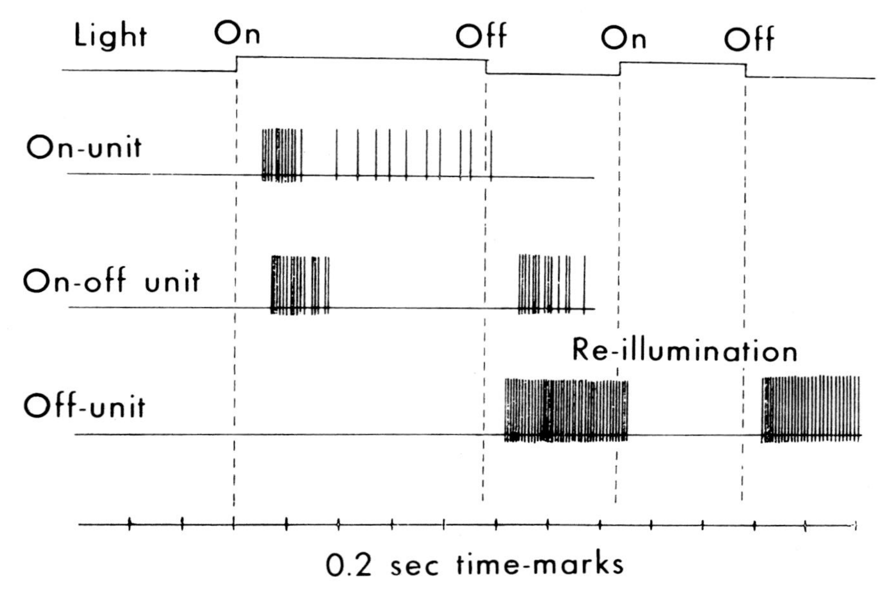

Fig. 3.13. Various types of responses recorded from optic nerve fibers of the frog (based on Figs. 1 and 5E of Hartline, 1938). The top line shows the extension in time of the light stimulus. The second line shows an "on" response, the third line an "on/off" response, and the fourth line an "off" response. The "off" response is inhibited by reillumination, as shown to the right of the recording on the fourth line. From Levick (1972).

was shut off represented a third response type (Fig. 3.13). These different responses led to the definition of "on-units," "on/off-units," and "off-units." Only one type of unit was recorded from each nerve fiber.

Under more favorable experimental conditions, Kuffler (1952, 1953) found that all three types of responses could be recorded from the receptive field of the same ganglion cell. Recordings made intracellularly from individual ganglion cells suggested that the receptive field consists of three different zones, the center, the surround, and the intermediate zone, (Fig. 3.14) with each zone characterized by a different type of response (Fig. 3.15). If illumination of the center evoked an on-response, illumination of the surround evoked an off-response, and illumination of the intermediate zone evoked an on/off-response. When the center responded with an off-response, illumination of the surround led to an on-response. The response of the surround was thus opposite that of the center.

The basis for this difference in the neural responses has not been revealed by the electrophysiological recordings. The effect of the sur-

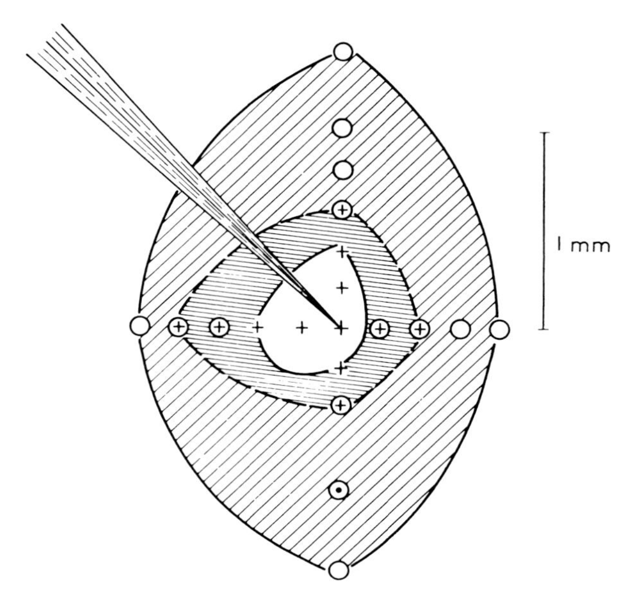

Fig. 3.14. Map of the receptive field of a ganglion cell in cat retina. "On" responses were evoked within the center and are indicated by crosses. "Off" responses, indicated by circles, characterize the surround, whereas "on/off' responses, cross within circle were obtained from an intermediary region. From Kuffler (1953).

round has been associated with the horizontal cells because of the large area from which responses from the surround can be evoked. The branchings of ganglion cells in the retina do not exceed 1.55 mm (Brown, 1965; Leicester and Stone, 1967), while the horizontal cells extend over considerably longer distances, with receptive fields measuring up to 10 mm in the fish retina (Gouras, 1960). Anatomically, the horizontal cells are therefore qualified for mediating the response of the surround.

Intracellular recordings from horizontal cells have shown that these cells are hyperpolarized by a light stimulus and that the hyperpolarization is graded and practically proportional to the logarithm of the light intensity until saturation. This was shown by the recordings of the S-potential of the L-type (Svaetichin, 1953). Numerous studies have confirmed this response type of horizontal cells in a variety of species.

That horizontal cells can affect the state of polarization of photoreceptors has been shown by Baylor and co-workers (1971) by current injection into horizontal cells and intracellular recording from cones in the turtle retina. Such feedback to photoreceptors from horizontal cells has also been shown by chemical blocking of the synaptic transmission between

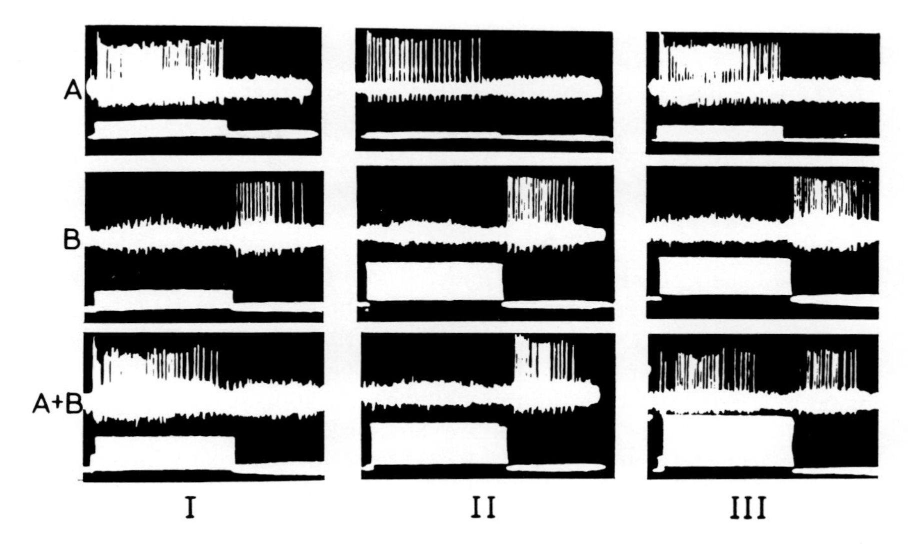

Fig. 3.15. Recordings from various parts of the receptive field of a ganglion cell in cat retina. Two spots of light were used to stimulate the photoreceptors, one projected onto the center and the other onto the surround. (A) Response to flashes of light on the center. (B) Response to flashes of light over the surround. (A + B) Responses after illumination of both center and surround. Flashing the spot over the center produces an on-response, whereas illuminating the surround generates an off-response. When both center and surround are illuminated simultaneously, the on-response is somewhat suppressed. If the relative strength of the light flashed over the center and over the surround is varied, the on-response can be completely suppressed when both center and surround are illuminated (column II). The recordings of column III show that the relative strength of the stimuli can be adjusted in such a way that simultaneous illumination of center and surround results in both an on- and an off-response, with both responses reduced relative to those when center and surround were illuminated separately. From Kuffler (1953).

horizontal cells and photoreceptors (Cervetto and MacNichol, 1972; Gerschenfeld and Piccolini, 1977).

The involvement of the horizontal cells in the response of the surround was shown by electrophysiological recordings by Maksimova (1969), who injected currents into horizontal cells while recording the responses of single ganglion cells. These responses were identical to those induced by stimulation by light. Four types of ganglion cell responses were recorded, including on-units, off-units, and on/off-units. When the injected current was changed from a hyperpolarizing to a depolarizing current, the opposite type of response was recorded, mimicking the center–surround antagonism.

The analysis of ganglion cell responses to current injection into horizontal cells was extended further by Naka and Nye (1970) and Naka (1971), who studied the catfish retina. Depolarization of a horizontal cell was shown to lead to a response similar to that induced by a spot of light projected on the center of the receptive field. Hyperpolarization of the horizontal cell gave rise to a response similar to that observed when the surround was stimulated.

To explain these types of responses one must know at which level they are established. That a current injected into the horizontal cells evokes the same responses as stimulation by light shows that the neural connections in the outer plexiform layer must contribute importantly to shaping the ganglion cell responses. This is further evidenced by the observations made in connection with the intracellular recordings from bipolar cells.

Such intracellular recordings are technically difficult and were first carried out successfully by Bortoff (1964) on the especially large bipolar cells of the mud puppy, *Necturus maculosus*. Since then a number of researchers have recorded from bipolar cells in several other species, all of which, however, have large bipolar cells (Byzov, 1966; Kaneko and Hashimoto, 1969; Werblin and Dowling, 1969; Kaneko, 1970; Matsumoto and Naka, 1972; Naka and Ohtsuka, 1975). The entire population of bipolar cell types has not been analyzed because even in the mud puppy retina there are bipolar cells too small for insertion of a microelectrode without damage preventing intracellular recording.

The receptive field of bipolar cells consists of a center and an antagonistic surround, showing that this type of organization of the ganglion cell receptive field is established at the bipolar cell level. The center of the bipolar cell receptive field measures 100–200 μm in diameter in the *Necturus* retina (Werblin and Dowling, 1969; Kaneko, 1973), which corresponds roughly to the extent of the dendritic arborizations of the *large* bipolar cells. The surround measures 1–1.5 mm in diameter (Kaneko, 1973).

Three types of bipolar cell responses have been recorded. The bipolar cells can respond with sustained hyperpolarization or sustained depolarization when the center is illuminated. Transient depolarization when the light stimulus is both turned on and shut off has also been recorded. These three types of responses correspond to the type 1, type 5, and type 9 responses deduced on the basis of the neural connections.

The recordings clearly reveal that in the depolarizing bipolar cells the sign of the potential change imposed by the photoreceptors has been reversed. This requires the participation of a neuron that can contribute to the reversal of the sign. Structurally, the horizontal cells are the only neurons that are in a position to do that, and the connections at the

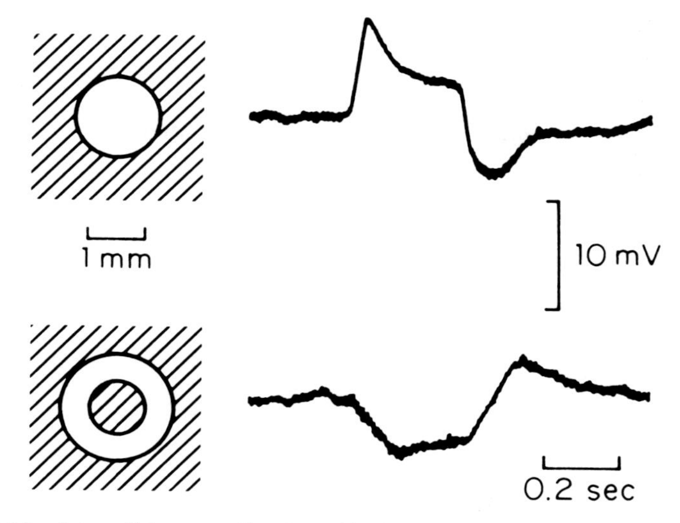

Fig. 3.16. Intracellular recording from bipolar cells (Kaneko, 1970). The tracing has been redrawn in such a way that hyperpolarization is shown by upward deflections to conform with the presentation of the polarity of the deduced potential changes.

synaptic ribbon complexes are the only connections that can account for a horizontal cell influence on the transmission between the photoreceptors and the bipolar cells strong enough to account for a sign reversal. The center–surround antagonism is also a case of sign reversal, and the horizontal cells are also in this case structurally the only neurons that can be involved in the reversal of the sign.

In the case of the hyperpolarizing bipolar cells, a light stimulus confined to the center evoked a sustained hyperpolarization that was reversed when the light was shut off. This corresponds to the type 1 response deduced above. When the light stimulus was confined to the surround, the hyperpolarizing bipolar cell was depolarized, in agreement with the type 3 response. When both the center and at least part of the surround were illuminated, the response became more complex. The hyperpolarization was reduced in amplitude after a certain delay, and when the light was turned off, the consequent depolarization of the bipolar cell overshot the baseline potential, in agreement with the type 2 response (Fig. 3.16).

That "part of the surround" should also be illuminated shows that the horizontal cells must be hyperpolarized to a certain level before they can exert this influence on the bipolar cell, in agreement with the deduction that the horizontal cell modulating influence is reduced and eventually vanishes when the photoreceptors are strongly depolarized at a low

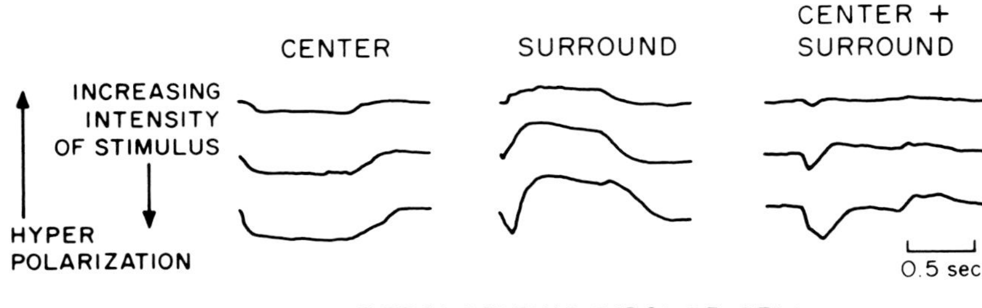

Fig. 3.17. Intracellular recordings from depolarizing bipolar cell (Naka and Ohtsuka, 1975). The polarity of the tracings has been reversed. The recordings clearly show that the on-response is evoked when the surround is illuminated. The inhibition of the on-response observed by Kuffler (illustrated in Fig. 3.15) which is evoked by the stimulation of the surround can therefore not occur at the level of the bipolar cells.

illuminance. This requirement explains the observations by Kaneko and Hashimoto (1969) that the bipolar cells respond with a sustained change in membrane potential when both the center and the surround are illuminated at low light intensities and that the response is composite at high light intensities.

Naka and Ohtsuka (1975) recorded a sustained depolarization of bipolar cells when the light stimulus was confined to the center, which is the type 5 response. When the stimulus was confined to the surround, the bipolar cell responded with transient depolarization, followed by sustained hyperpolarization (Fig. 3.17). This response, particularly pronounced at high stimulus intensities, corresponds to type 6.

However, the hyperpolarization following the initial depolarization is more extensive than the change deduced for a type 6 response. Obviously, only qualitative, not quantitave, aspects of changes in the membrane potential can be deduced.

The fact that when the light stimulus is confined to the surround, the depolarization of the bipolar cell is obvious only at a high stimulus intensity indicates that the local input from photoreceptors to the horizontal cells at the synaptic ribbon complexes is so strong that it requires strong input from the surround to break the hyperpolarization block.

Naka and Ohtsuka found that illumination of both center and surround evokes a transient depolarization when on and a slight transient hyperpolarization or no response when off. This corresponds to type 7 of the deduced responses.

The responses of the depolarizing bipolar cells thus agree rather well with those deduced for these cells. Naka and Ohtsuka concluded that the transient depolarization when high-intensity light over the surround was turned on was not caused by stray light. They offered no explanation for the response.

This response is of a particular interest because it clearly shows that the surround of the depolarizing bipolar cells, in agreement with the deduction, does not inhibit the bipolar cell response. It shows that the inhibition of ganglion cell responses that can be generated by the stimulation of the ganglion cell surround (Kuffler, 1953) does not reflect bipolar cell responses. This inhibition must therefore be imposed by neurons in the inner plexiform layer, where the amacrine cells are the only neurons that anatomically are capable of mediating inhibition over long distances laterally.

The structural analysis revealed that bipolar cells connected to synaptic ribbon complexes in most cases are connected to both mixed and nonmixed complexes, with mixed complexes dominating numerically. The deduced responses involve a combination of the on-response with the depolarization partially sustained at a lower amplitude during the light stimulus. The connections at unmixed complexes would contribute to the latter aspect of the response. This deduction can be related to Hartline's (1938) observation that the nerve fibers from which an on-response was recorded responded with a high-frequency on-discharge followed by a low-frequency discharge during the light stimulus.

Kaneko and Hashimoto (1969) showed that the off-response of on/off-bipolar cells is suppressed during dark adaptation. This observation agrees with the deduction that in the dark-adapted state the hyperpolarizing pressure exerted by the large horizontal cells is maintained at a maximum, while the small horizontal cells exert a lower hyperpolarizing pressure on the bipolar cell endings. The combined hyperpolarizing pressure blocks the transmission of signals between the photoreceptors and the bipolar cell. An on-response can, however, be generated because the light stimulus reduces the hyperpolarizing pressure exerted by the small horizontal cells, opening the gate for transmission of a signal from the photoreceptors to the bipolar cell. No such modulating action affects the off-response.

In our deductions we ascribed an important role to the horizontal cells in the shaping of the bipolar cell response. The observations of Naka (1971) justify this basis for our conclusions. While injecting a current into a horizontal cell, Naka simultaneously recorded the responses of a ganglion cell and of a cell in the inner plexiform layer that was most likely a bipolar cell. The presumed bipolar cell responded with slow changes in its membrane potential, while the ganglion cell generated bursts of spike potentials. Whether an on-response or an off-response was recorded depended on the sign of the polarizing current injected into the horizontal cell. Because the horizontal cells exert this influence, the response of the bipolar cell must be determined by a balance between the inputs from the photoreceptors and from the horizontal cells, in agreement with our deductions.

It has been shown that the influence of the surround on the center is not noticeable in the dark-adapted retina (Barlow *et al.*, 1957). This agrees with our deduction that the horizontal cell modulating influence decreases and eventually vanishes when the horizontal cells approach and reach maximal depolarization in the dark-adapted retina.

The electrophysiological recordings have also revealed one type of neuron in the inner nuclear layer that responds with depolarization when either the center or the surround is illuminated (Kaneko and Hashimoto, 1969), agreeing with type 8 of the deduced responses. According to our deductions, such a unit would be a depolarizing bipolar cell (L-bipolar cell) that contacts only large horizontal cells at synaptic ribbon complexes such as one of the ten bipolar cells that contacted terminal 1.

Oscillatory responses have been recorded by Normann and Pochobradsky (1976) from rods and horizontal cells. The oscillations disappeared after perfusion with aspartate, which blocks the synaptic transmission between photoreceptors and horizontal cells. This shows that the oscillations are due to photoreceptor–horizontal cell interaction.

We thus can conclude that the types of responses deduced on the basis of the observed structural relationships among the neurons agree well with the observed responses. That this also applies to rather complex reponse patterns justifies the conclusion that the deductions furnish a reasonable explanation for the observed responses of bipolar cells.

The observed center off and surround on responses at the ganglion cell level can be explained by the deduced and observed responses of hyperpolarizing bipolar cells. This response pattern can therefore be accounted for by neural interactions at the level of the outer plexiform layer.

The responses from the intermediate zone of the ganglion cell receptive field can be explained by the light stimulating the center of some bipolar cells and only the surround of others.

The combination of center on and surround off responses does not correspond to any response pattern of the bipolar cell. Center on-responses can be generated by depolarizing bipolar cells but no off responses can be generated from the surround of these cells. Therefore the center on and surround off combination must involve interference from neurons outside the outer plexiform layers, and it seems highly likely that the inner plexiform layer also contributes to these responses of the ganglion cells.

25. The Bipolar Cell Response to a "Stationary" Image

The bipolar cells that respond with on-, off-, or on/off-responses transmit signals regarding changes in the intensity of light falling on their receptive fields. Between these changes the membrane potential is maintained at a baseline potential. On-bipolar cells connected to both mixed and nonmixed synaptic ribbon complexes are exceptions, and together with the L-bipolar cells they continuously transmit information regarding the average illuminance over their receptive fields.

When an object is viewed steadily for a certain time, the boundaries of the object and of any patterns on the object generate on- and off-responses as a result of the involuntary eye movements that make these boundaries oscillate over a certain limited area of the retina, continuously changing the intensity of the light projected onto the individual photoreceptors within this area. The L-bipolar cells connected to photoreceptors located away from the boundaries are not exposed to the same abrupt changes, and L-bipolar cells located at the boundaries vary the amplitude of their sustained responses according to an averaging of the illuminance over the entire receptive field.

We can then predict that it is the on-, off-, and on/off-bipolar cells that account for the transmission of information regarding boundaries in the image pattern projected onto the retina. If we eliminate eye movements when viewing an object, we eliminate the stimulus for these bipolar cells, and the boundaries should vanish. Only the luminosity responses should still be evoked. Ditchburn and Ginsberg, (1952) and Riggs *et al.* (1953) showed experimentally that this is the case.

26. The Surround Is Not a Functionally Unique Part of the Bipolar Cell Receptive Field

The responses that can be generated by changing the intensity of the light over the surround are not unique to the surround. Those responses are evoked by the horizontal cells and the responses will be identical whether they are generated from the surround or from the center. The only difference between these two alternatives is that in the latter case the responses are the consequence of the horizontal cell influence being mixed with the influence exerted directly by the photoreceptors.

The explanation for the bipolar cell surround appearing to be an area from which unique antagonistic responses can be evoked is the horizontal cells' requiring a certain minimum input from photoreceptors to modulate the bipolar cell responses. When the entire receptive field is illuminated uniformly, the horizontal cells receive an input from photoreceptors in the surround that exceeds by far the corresponding input within the center because the surround covers a considerably larger area than does the center. However, the same type of responses from the horizontal cells should be evoked if the smaller area of the center is compensated for by an increase in the stimulus intensity.

When in electrophysiological experiments the stimulation of the center with high-intensity light evokes responses in which the response typical of stimulation of the center is mixed with a response typical of stimulation of the surround, it has been assumed that the surround has been stimulated by stray light or that the illuminated area has included "part of the surround."

Obviously these observations can be explained by the input to the horizontal cells within the center being large enough for the horizontal cells to modulate the bipolar cell responses. Consequently, there is no reason to assume that the surround has been involved in the stimulation, and there is no reason to ascribe any unique properties to the surround.

27. Are Structural Connections between Neurons Also Functional Connections?

We pursued the structural analysis of the outer plexiform layer on the premise that the function of the observed circuitry could be deduced from the pattern of neural connections. This premise is based on the assumption that these connections are functional and represent synapses. We also assumed that conclusions regarding the physiology of a neural center can be drawn from structural relationships and that the boundary between morphology and physiology has been abolished because the structure can be translated directly into function.

That the boundary between morphology and physiology can be abolished is illustrated clearly by earlier studies of the nervous system. The discovery of the neuron as a structural unit of the nervous system is one example of structure being translated directly into function. The connections established by processes between neurons was translated into function by the interpretation that these connections allow the neurons to communicate. The structural analysis of the nervous system has, to a considerable extent, involved the establishment of neural connections, and physiological analysis has confirmed that such connections allow communication within the nervous system.

The analysis of the circuitry of the outer plexiform layer revealed neural connections within this layer that are established by special processes endings of which contact other neurons. It should be emphasized that the connections considered during this study are established by special end branches of bipolar cell dendrites and by horizontal cell processes that extend for some distance after branching from the neural process to reach the site of contact.

One difference between the information collected during this study and that obtained by Ramon y Cajal and followers regards the dimensions of the processes. The electron microscope eliminated the limitations that were imposed by the resolving power of the light microscope. It is highly improbable that the functional significance of neural processes changes with their dimensions and that the minimum size of a functional connection coincides with the limit at which metal-impregnated processes can be observed in the light microscope.

That the observed connections are functional is made highly probable by several circumstances. First, the connections occur according to certain repeating patterns. It is highly unlikely that nonsensical connections would follow such patterns; more likely they would be randomly arranged.

Second, the observed connections are the only ones between photoreceptors on the one side and horizontal cells and bipolar cells on the other. They are also the only connections between horizontal cells and bipolar cells.

Third, in most cases the connections are associated with accumulations of synaptic vesicles or opaque granules. However, the synaptic vesicles in certain end branches, such as the horizontal cell endings at synaptic ribbon complexes, are particularly susceptible to destruction during the preparatory procedure, as shown by the great variation in the number of synaptic vesicles in these endings in different preparations and also in different synaptic ribbon complexes in the same preparation. An absence of synaptic vesicles in the embedded material therefore does not necessarily mean that there were no synaptic vesicles present in the intact living tissue.

Other structural specializations, such as opacities at the neural membrane, differences in the thickness of the membrane, and opaque material in the gap separating the membranes of the contacting processes, characterize many of the neural connections. The functional significance of these specializations is unknown.

Adoption of certain morphological criteria for a functional connection would be justifiable if we knew the functional significance of these criteria through an analysis of synaptic transmission correlating the structural features to specific aspects of signal transmission. If such a correlation revealed that at the level of dimensions of these specializations transmission requires a particular structural organization, we would be justified in applying these morphological criteria. In the absence of any such correlation these criteria are based only on arbitrary, subjective judgment.

One way to identify a functional connection is to demonstrate the presence of a receptor protein in the postsynaptic membrane. This is, however, not possible because the preparatory technique destroys the specimen's molecular structure. The basic properties of the pre- and postsynaptic membranes that associate their structure with their function are to be found at the molecular level while the specializations associated with synapses so far are of much coarser dimensions.

Nevertheless, structural criteria for a functional connection have been applied. It has been claimed that the presence of synaptic vesicles is not sufficient to indicate that a connection is functional. The vesicles must be clustered. This conclusion is presumably based on clustered vesicles looking like a more specialized structure than nonclustered vesicles. This is obviously a purely subjective way to evaluate the observed structure.

Typically, clustered synaptic vesicles are present in synaptic knobs, where the number of vesicles is small in comparison to the volume of the ending. They are usually not clustered in the synaptic terminals of the photoreceptors, where the vesicles are numerous, but the neural connection there must still be functional.

A most strange interpretation of the structure of synaptic ribbon complexes involves confining the synaptic contact area between horizontal cells and the photoreceptor to the septum containing the synaptic ribbon. Presumably the often regular arrangement of synaptic vesicles at the ribbon surface has conveyed the impression of a clustering of vesicles, and the ribbon structure contributes further to make the connection appear special enough to be functional!

The clustering of vesicles can be interpreted as reflecting an arrangement by which transmitter release can occur locally by the anchoring of the vesicles at the presynaptic membrane region. Obviously, the clustering of the vesicles cannot be explained unless the vesicles are anchored to this region because without anchoring they would disperse uniformly in the cytoplasm of the synaptic knobs. For instance, the vesicles may be anchored in a gelled region of the cytoplasm at the presynaptic membrane. Fibrous protein might well be responsible for the gelled state of that cytoplasm. During fixation and dehydration of the tissue this gel is likely to shrink, exaggerating the clustering of the vesicles and concentrating the fibrous protein at the presynaptic membrane into a cytoplasmic density. In fact, it is highly likely that artifacts of preparation greatly modify the appearance of neural connections.

The correlation between the presence of vesicles and synaptic connections is well established. However, an apparent absence of synaptic vesicles does not prove that the connection is nonfunctional because the vesicles may have been destroyed during the preparatory procedure. In the material analyzed in this study there were connections at which synaptic vesicles were absent. The fact that synaptic vesicles were present at other connections of the same kind, both in the same specimen and in other specimens, makes it likely that the absence was due to poor preservation of the vesicles and not to a real absence *in vivo*.

If the end branches observed during the analysis of the outer plexiform layer do not establish functional connections, what other explanations for them might we propose? Perhaps they have a space-filling function. Perhaps they reflect a deficiency in the way the nervous system is assembled, leading to the neurons extending a large number of nonsensical processes. In both cases these processes would just happen to be organized according to systematic patterns. A third possibility is that the processes are mere decorative contributions of the neurons and that the systematic patterns reflect neuronal aesthetical requirements.

The decisive factor in an evaluation of the functional role of the observed neural connections is whether the deductions made on the basis of these connections assist us in developing an understanding of the way a neural center processes information. If the processes are nonsensical they would constitute a poor basis for any such deduction.

Chapter 4: The Outer Plexiform Layer is a Major Center for Information Processing

1. The Subsynaptic Neuropil: Miniaturized Circuits

Let us now extend the analysis to the part of the outer plexiform layer vitread to the base of the photoreceptor terminal. Surprisingly, there is a rather complex system of neural connections at this level, where it had been expected that bipolar cell and horizontal cell dendrites would just pass through to make contact with the photoreceptors.

The complexity of the structure of this part of the outer plexiform layer is clearly shown in Figs. 4.1–4.4. Detailed analysis of the intertwined neural processes (Sjöstrand, 1974, 1976) revealed that connections are established between these processes by special side branches extending from one process and ending in a contact with another process. The dendrites from which such side branches originate either pass through the region contacting other processes with side branches before ending in contact with the photoreceptor terminal or else contact neural processes vitread to the terminal without making any contact with the terminal itself. The neural processes are closely packed and there are no Müller's cell processes between them.

It should be emphasized that the contacts are made by special branches that end at the site of the contact and that the contacts do not involve just close apposition of the processes as they pass through the region to the terminal. With no Müller's cell processes present in this region there are obviously extensive contacts among the closely packed processes. An analysis of these latter contacts revealed that they are random, with the exception of two pairs of bipolar cells.

At these latter contacts the neural membrane is particularly thin and the membranes are separated by a space narrower than the space separating other processes. These two pairs of connections were not interpreted as being synaptic connections, but they may play a role in synchronizing the signals transmitted by the two paired bipolar cells.

The connections of these two pairs of bipolar cell dendrites are nonrandom. The two bipolar cells of each pair were connected to the same combination of horizontal cells. Both bipolar cells of one pair were thus connected to horizontal cells approaching the terminal from north and west, while both the bipolar cells of the other pair were connected to horizontal cells approaching from south and west. All four bipolar cells were on/off-bipolar cells. While one bipolar cell within each pair was connected to both large and small horizontal cells, the other bipolar cell was connected to only large horizontal cells. One bipolar cell in each pair was therefore an L-bipolar cell. The close association of the paired bipolar cells is shown in Fig. 1.32.

The special side branches that contribute to the neural connections contain accumulations of synaptic vesicles or small opaque granules at the site of contact (Fig. 2.20).

The extent of neural connections is shown by the number of neurons involved in such contacts. For instance, at terminal 1, twelve neurons, mostly bipolar cells, contributed to such connections. The special side branches from the processes of these twelve neurons established more than 40 contacts at terminal 1. At terminal 2, ten neurons, five bipolar cells, and five horizontal cells contacted branches of bipolar cell 1 alone at ten different sites.

The absence of Müller's cell processes contributes to the compactness of the circuitry as do the small dimensions of the processes. Part of the thinnest processes measured only about 600 Å in diameter, and the average thickness of the neural branches was only 1200 Å. These small dimensions made it possible for the contact to be established by very short processes. The presence of synaptic vesicles in the short end processes (Fig. 2.20) shows that the connections in this region are synaptic.

Figure 2.23 gives an account of neural connections at the base of terminal 1. The neural connections were more extensive at terminal 2, but although they were traced and reconstructed three-dimensionally, the detailed analysis of the connections has been confined so far to those involving bipolar cell 1 at this terminal.

Part of this circuitry is located in a deep recess formed by the concavity of the basal surface of the terminals. This is shown, for instance, in Fig. 1.24, where an accumulation of neural profiles occupies a central area surrounded by the terminal. The depth of this recess can be determined from the number of sections in which this central area is present. It is about 1 μm deep. However, the region in which such interneural connections are present extends vitread 1.5 to 2 μm further.

Fig. 4.1. Anatomical model of terminal 1 after removal of most of the processes in the common neuropil. What remains is most of the subsynaptic neuropil, bipolar cell processes to the terminal, and the synaptic ribbon complexes. The latter complexes surround a large part of the subsynaptic neuropil located in the concavity of the basal surface of the terminal. Arrow points to one collateral of bipolar cell 1.

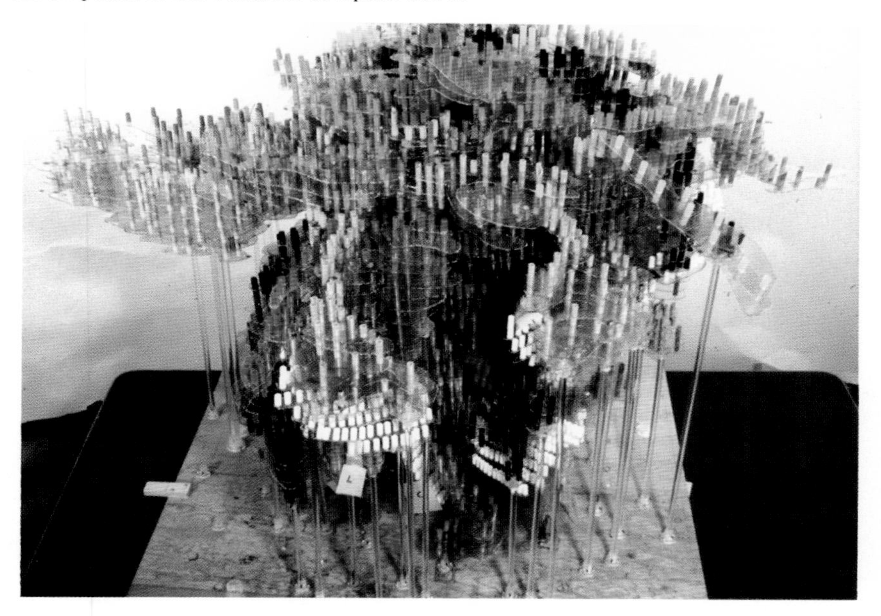

Fig. 4.2. Close view of part of the three-dimensional anatomical model, exposing the middle of the subsynaptic neuropil after a large part of it has been removed. The subsynaptic neuropil extends down between the synaptic ribbon complexes, which can be identified by the white synaptic ribbons. From Sjöstrand (1974).

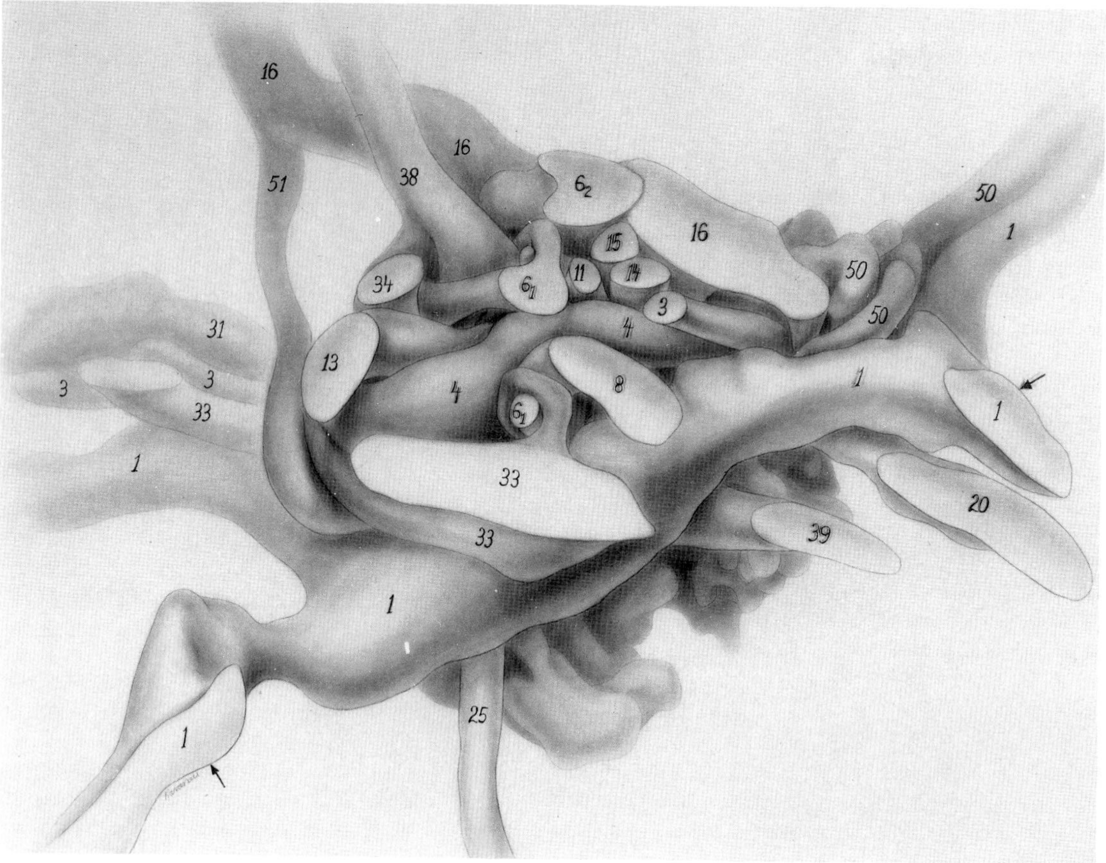

Fig. 4.3. Drawing of the model exposing part of the subsynaptic neuropil. Drawing by Mrs. Hermine Kavanau. From Sjöstrand (1974).

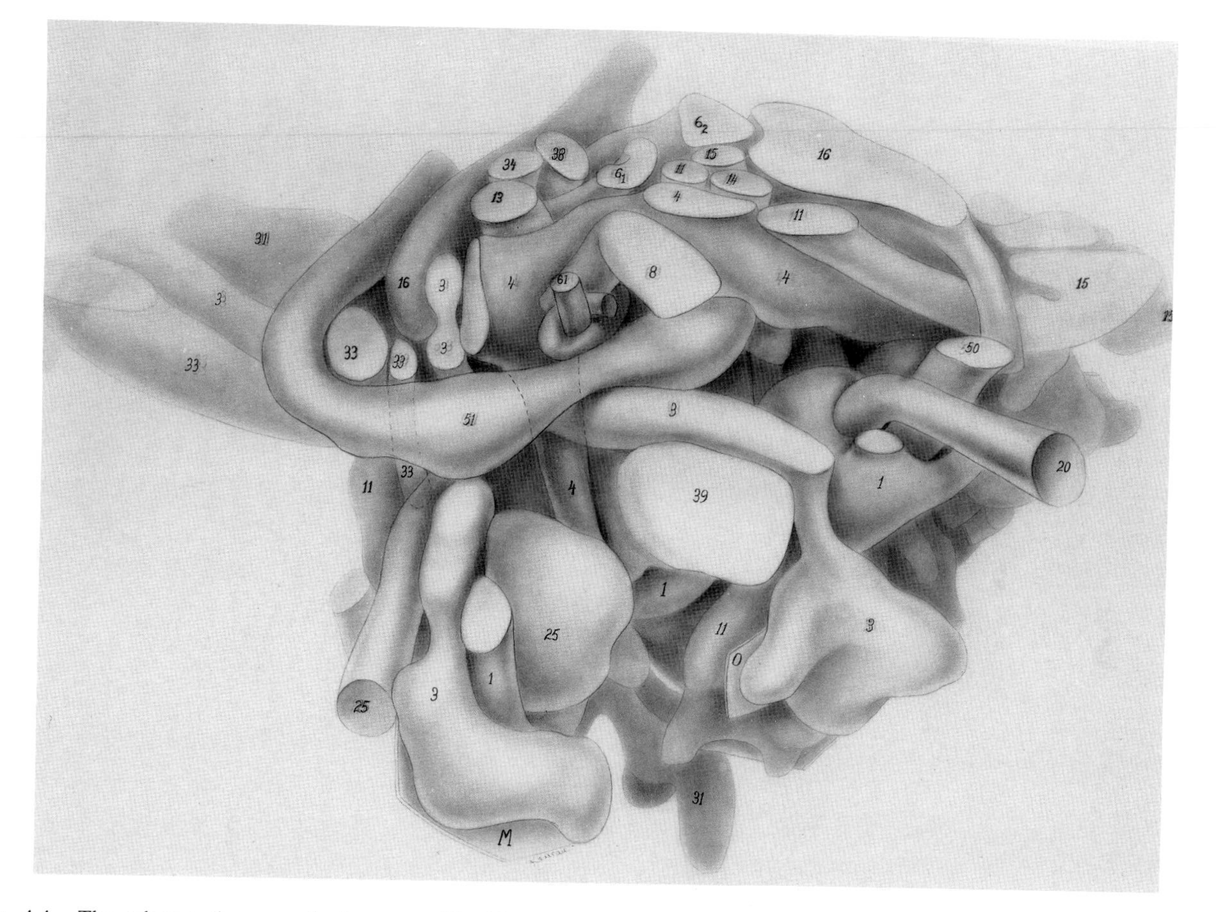

Fig. 4.4. The subsynaptic neuropil after a considerable part has been removed to reveal the part located predominantly between the the synaptic ribbon complexes. Drawing by Mrs. Hermine Kavanau. From Sjöstrand (1974).

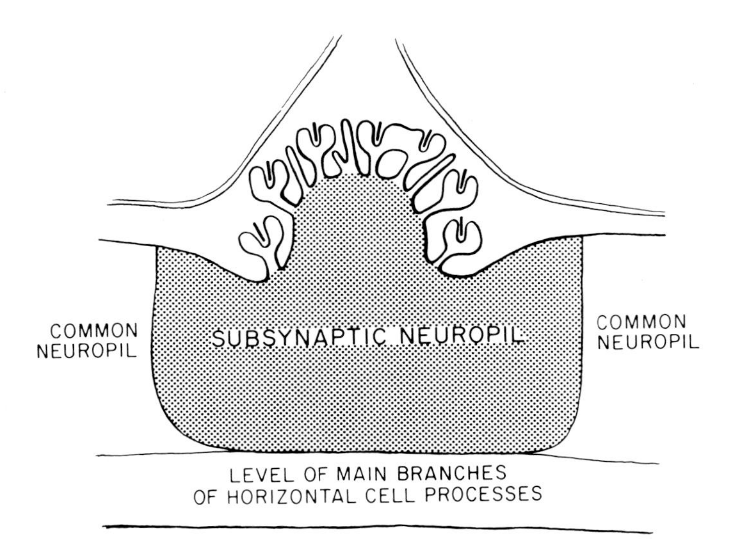

Fig. 4.5. Schematic representation of the extension of the subsynaptic neuropil at cone terminals.

The average diameter of the processes in the inner plexiform layer exceeds that of the processes at the photoreceptor terminals by a factor of 2 to 3. The difference in size between the latter processes and those in brain centers might be even greater. The circuitry at the base of each terminal thus may be considered to be miniaturized. This region of miniaturized circuits will be referred to as the *subsynaptic* or *subterminal neuropil* to distinguish it from the rest of the outer plexiform layer, *the common neuropil*. The extension of the subsynaptic neuropil is illustrated schematically in Fig. 4.5. The discovery of this circuitry was most surprising. It shows that conditions for rather complex neural interactions at the level of the photoreceptors are fulfilled. Each photoreceptor is thus associated individually with a circuitry, allowing direct neural interaction as signals are transmitted from the photoreceptor.

Let us examine an example of the complex circuitry in the subsynaptic neuropil, the connection of bipolar cell 1 in the subsynaptic neuropils of terminals 1 and 2 (Figs. 4.6A,B). These contacts involved special end branches contributed either by bipolar cell 1 or by other neurons. In some cases the contact was made by one process passing by a bipolar cell 1 branch. When specialized structures such as a widening of the process, a specialized membrane structure, *and* synaptic vesicles were present at the site of contact, these in-passing contacts were considered to be functional. In no case was a contact relationship between two neurons confined to such in-passings contacts but always involved in addition connections made by the ends of special end branches.

Bipolar cell 1 was contacted by a total of nine bipolar cells at the two terminals, four at terminal 1 and five at terminal 2. At the latter terminal five horizontal cells were also involved in such contacts. In most cases, the neurons made multiple contacts, bringing the number of contacts to 23 at 10 sites at terminal 2.

Just as at terminal 1, the connections of bipolar cell 1 in the subsynaptic neuropil at terminal 2 involved other bipolar cells. However, a contacting bipolar cell was in most cases joined by one horizontal cell with the bipolar cell and the horizontal cell in mutual contact at the site where they both contacted bipolar cell 1. Thus a triad connection was established with a pair of bipolar cells being contacted by one horizontal cell. In some cases a second horizontal cell was involved in the multineural connection. At only one site was bipolar cell 1 contacted by a pair of bipolar cells without a horizontal cell being involved. In this case the three bipolar cells formed a triad connection. One of these bipolar cells was an efferent bipolar cell.

Of the horizontal cells, three were small and four were large. All small horizontal cell endings were contributed by major processes that approached the terminals from north. No branches of the main processes of the small horizontal cells contacted photoreceptors south of the terminals.

Either large or small horizontal cells contributed endings to the connections between bipolar cell 1 and other bipolar cells. Three bipolar cells that contacted bipolar cell 1 made multiple connections and were in contact with both large and small horizontal cells, though at different sites. Of the other bipolar cells that contacted bipolar cell 1, one contacted a single large horizontal cell at two different sites.

The combination of bipolar cell and horizontal cell endings at the ten sites of contact with bipolar cell 1 at terminal 2 are reminiscent of the pattern of connections at the synaptic ribbon complexes. Most bipolar cells contacted endings of both small and large horizontal cells, and all endings of small horizontal cells were endings of branches that originated from processes that approached the terminal from north, as did the processes that contributed endings to the synaptic ribbon complexes. Furthermore, none of the main processes of the small horizontal cells contributed any end branches contacting photoreceptors located south of terminals 1 and 2. This means that interaction between bipolar cell 1 and most other bipolar cells in the subsynaptic neuropil is under a modulating influence exerted by horizontal cells, a situation somewhat resembling that at the connections to the terminals.

This analysis of the contacts of bipolar cell 1 in the subsynaptic neuropils clearly shows that this region can accomodate rather complex circuitry within a very limited volume of neural tissue. It also shows that the connections of neurons in this region involve predominantly connections between bipolar cells.

The synaptic ribbon complexes and a large part of the subsynaptic neuropil are not included in what is referred to as the outer plexiform layer, a term based only on gross structural features that can be observed at the resolution of the light microscope. The size of the terminals prevents them from forming a single layer. Instead, they are staggered in a way similar to the staggered arrangement of the nuclei of the photoreceptors in the outer nuclear layer. The fact that each photoreceptor is associated with a subsynaptic neuropil in addition to synaptic ribbon complexes therefore makes the miniaturized circuitry of the outer plexiform layer extend far beyond the recognized sclerad boundary of this layer. If we include this part of the circuitry in the outer plexiform layer, the thickness of this layer is doubled.

The "outer" and "inner" plexiform layers are defined on the basis of the light microscopic appearance of the retina, which does not reveal the circuitry described above. From both a morphological and a functional point of view it seems justifiable to assign all the circuitry revealed by electron microscopy at the level of the photoreceptors to the same layer of the retina. When so defined, this layer extends far beyond the sclerad boundary of the outer plexiform layer. To change the old terminology

---

Fig. 4.6. Connections of bipolar cell 1 in the subsynaptic neuropil at terminal 1 (A) and terminal 2 (B). The subsynaptic neuropil at terminal 2 was analyzed more extensively with respect to horizontal cell connections. (B) Bipolar cell 1 was connected to processes from other bipolar cells at 10 sites; these sites are indicated in the first column on the right. The other three columns show the combinations of bipolar cell and horizontal cell connections and the numbers of large (LHC) and small (SHC) horizontal cells. Note the presence of synaptic vesicles in widened parts of the process of bipolar cell 1 at sites 2, 4, and 6 and in a side branch at site 7. Because synaptic vesicles are present at the connections of bipolar cell 1, this bipolar cell is the presynaptic neuron relative to bipolar cell 41_2 and 22_2, and it is the postsynaptic neuron relative to bipolar cells 29_2 and 85_2 and the efferent bipolar cell 15_2. Its relationship to bipolar cell 15_2 agrees with that to the efferent bipolar cell 14 at terminal 1. The connection to bipolar cell 22_2 may be a reciprocal synaptic connection at site 4. Bipolar cells 15_2, 41_2, and 85_2 are connected to both large and small horizontal cells at different sites, whereas bipolar cell 22_2 is connected only to the same large horizontal cell at two sites. Bipolar cell 29_2 is not connected to any horizontal cells.

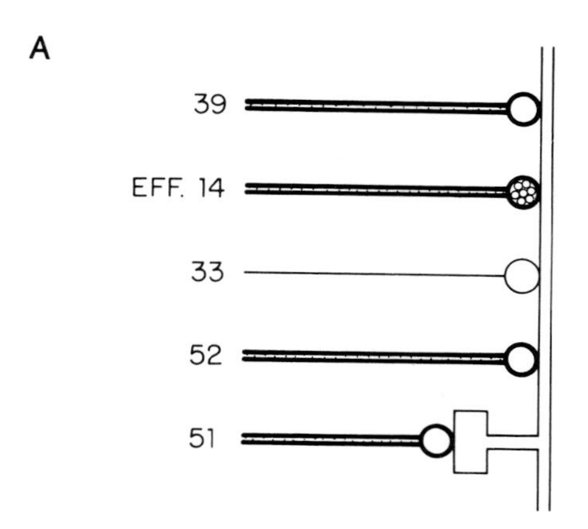
A
39
EFF. 14
33
52
51

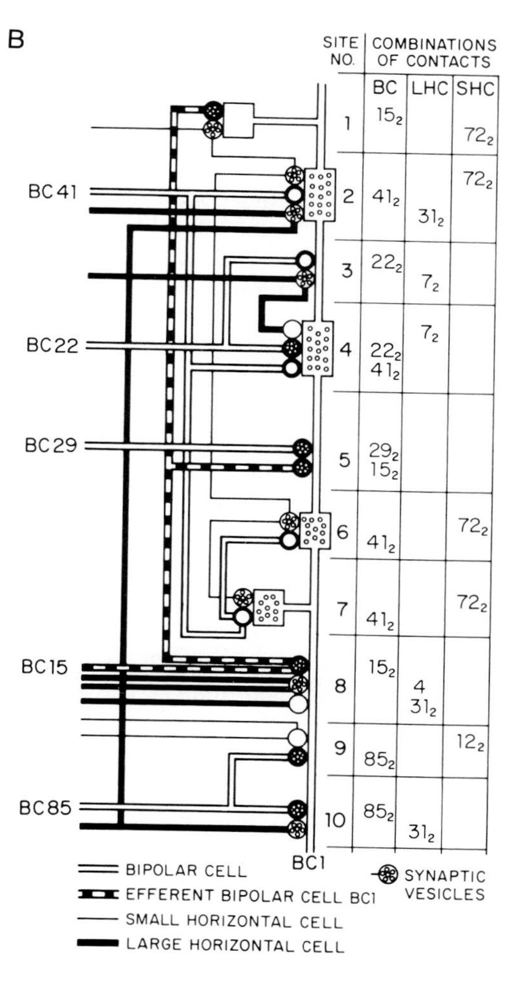
B
SITE NO.
COMBINATIONS OF CONTACTS
BC
LHC
SHC
1 15_2 72_2
2 41_2 31_2 72_2
3 22_2 7_2
4 22_2 41_2 7_2
5 29_2 15_2
6 41_2 72_2
7 41_2 72_2
8 15_2 4 31_2
9 85_2 12_2
10 85_2 31_2
BC41
BC22
BC29
BC15
BC85
BC1
BIPOLAR CELL
EFFERENT BIPOLAR CELL BC1
SMALL HORIZONTAL CELL
LARGE HORIZONTAL CELL
SYNAPTIC VESICLES

appears too radical, however. We therefore propose a new terminology that can be used together with the old, the two terminologies referring to two different aspects of the structure of the retina. The region in the retina containing neural connections of photoreceptors, bipolar cells, and horizontal cells and the processes contributing these connections we will refer to as the *outer circuitry layer* to distinguish it from the light microscopic definition of the outer plexiform layer.

2. The Role of the Outer Circuitry Layer in Information Processing

The thickness of the outer plexiform layer relative to that of the inner plexiform layer has led to the conclusion that the inner, not the outer, layer is involved in information processing. As explained in the preceding section, the difference in thickness is greatly reduced if we include in our comparison the parts of the circuitry that are located at the photoreceptor terminals. The inner plexiform layer is then only twice as thick as the outer circuitry layer. Still, the comparison is misleading because it does not take into account several features that should be considered in an evaluation of the relative capacities of the two layers for information processing. This capacity is determined by the number of neural connections that can be accommodated in the two layers, which in turn depends on the average dimensions of the neural processes. With the average thickness of the neural processes in the inner plexiform layer exceeding that of neural processes in the subsynaptic neuropils by a factor of 2 to 3, a proper correction is likely to make the outer circuitry layer exceed the inner plexiform layer in capacity to process information. The corrected thickness of the outer circuitry layer will be determined by the correction factor to the third power because only the thickness of the layers can vary.

A fair comparison also requires that the extent to which Müller's cells contribute to the volume of the layers be considered. In the inner plexiform layer the Müller's cells account for a considerable part of the volume, while in the subsynaptic neuropils no Müller's cell processes are present. The Müller's cells also form wide columns extending across the inner plexiform layer, occupying a considerably larger volume in this layer than in the outer plexiform layer. The axons of the bipolar cells occupy a large volume in the inner plexiform layer. These axons contribute the neural input to the inner plexiform layer. The corresponding input to the outer plexiform layer is located at the photoreceptor terminals outside the outer plexiform layer.

We conclude that the outer circuitry layer has a high capacity to process information and that its capacity may well exceed that of the inner plexiform layer. In this manner a large part of the information processing occurs at a resolution that corresponds to that of the information transmitted by the photoreceptors. The discovery of the circuitry at this level leads to the conclusion that no analysis of the function of the retina can be meaningful unless it includes the neural interactions in this part of the retina.

3. The Connections of Bipolar Cell 1 in the Common Neuropil

The common neuropil refers to the part of the outer circuitry layer vitread to the subsynaptic neuropil and extending sclerad between the subsynaptic neuropils of adjacent photoreceptors. The main branches of the horizontal cells are located in the vitread part of the common neuropil, and columns of Müller's cells mix with neural processes in this part.

Of the 18 bipolar cells that contacted photoreceptors 1 and 2, there was only one, bipolar cell 1, that sent branches out from the subsynaptic neuropil into the common neuropil, and it did so at both terminals 1 and 2. None of the other 17 bipolar cells connected to terminals 1 and 2 contributed any such branches at either terminal. The neural connections of bipolar cell 1 in the common neuropil are therefore a special feature.

The end branches that extended into the common neuropil will be referred to as *collaterals*. Bipolar cell 1 sent off four collaterals at terminal 2 and two at terminal 1. These processes differed from other end branches by their greater thickness. They thus were 0.30 to 0.33 μm thick as compared to 0.20 μm for the branches to the terminals before these branches split up into end branches. The latter were only 0.06 to 0.15 μm thick. Figure 4.7 shows a section that passes through three collaterals of bipolar cell 1 at terminal 2.

The analysis of the connections of these collaterals at terminal 1 and 2 revealed that in contrast to the situation in the subsynaptic neuropils all contacts except one involved horizontal cells. As in the synaptic ribbon complexes, the horizontal cell endings at the collaterals were paired and involved both large and small horizontal cells. The pairs were contributed by one small and one large horizontal cell with only one exception, in which at one site two small horizontal cells contributed the endings.

Figure 4.8 gives an account of the contacts of the collaterals of bipolar cell 1 at terminals 1 and 2. Including both terminals, 11 horizontal cells contacted bipolar cell 1 at its collaterals. Eight of these were small horizontal cells. All processes of these eight small horizontal cells approached the terminals from north and none contacted any photoreceptor located south of the terminals.

It is quite clear that the contacts between the collaterals and the

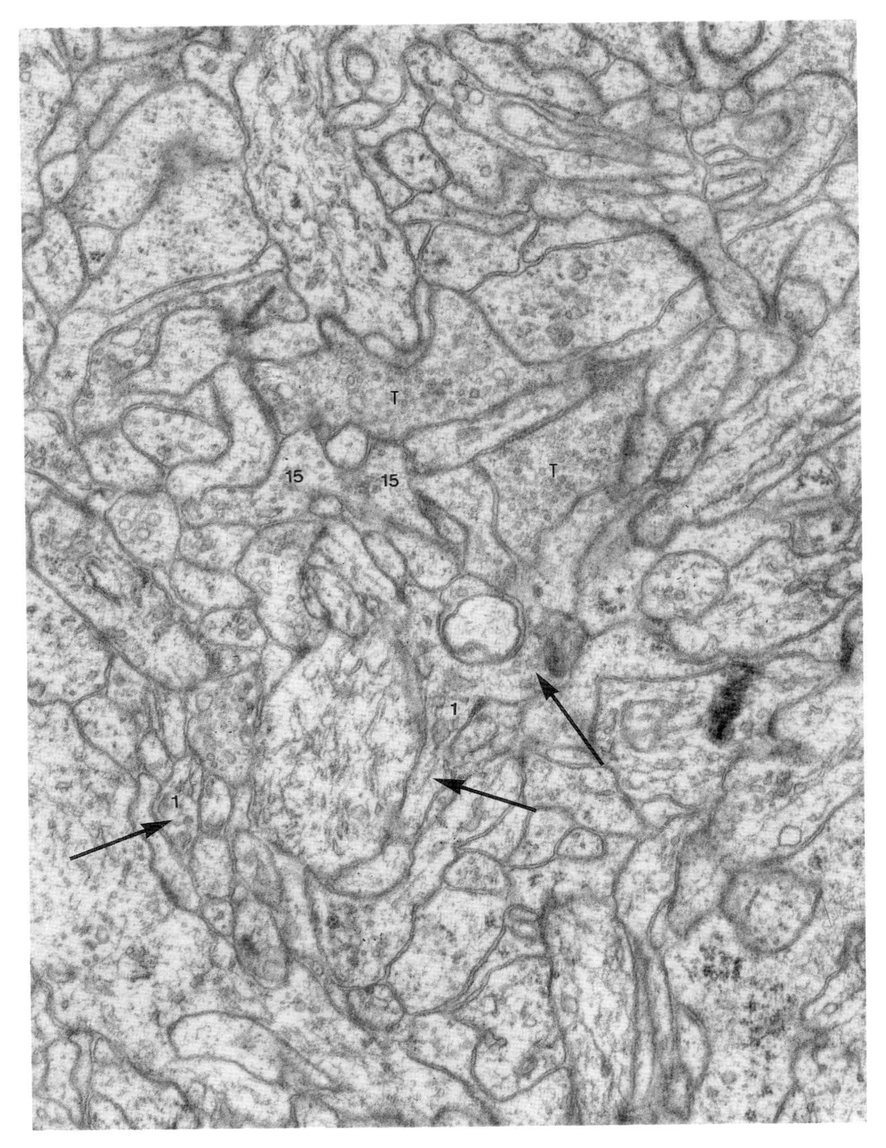

Fig. 4.7. Collaterals of bipolar cell 1 at terminal 2 (arrows). T, terminal 2. 25,000×. From Sjöstrand (1978).

horizontal cells show a definite pattern of horizontal cell connections that is similar to the pattern in synaptic ribbon complexes. One difference between the contacts at the synaptic ribbon complexes and those at the

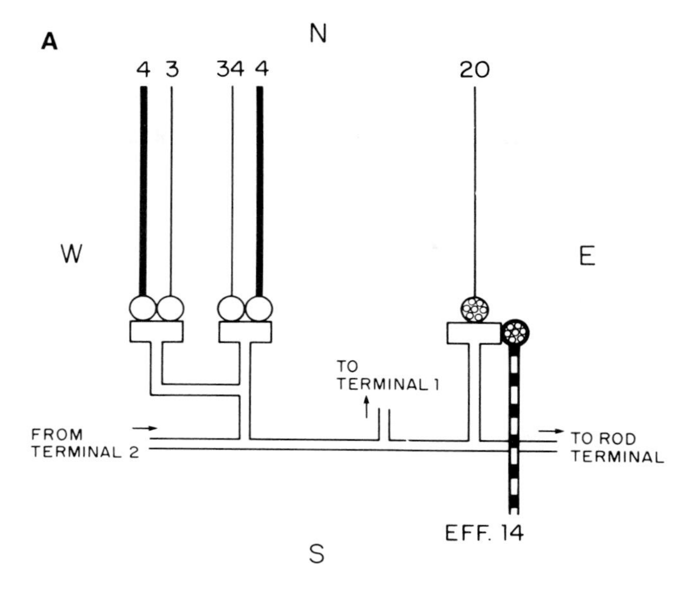

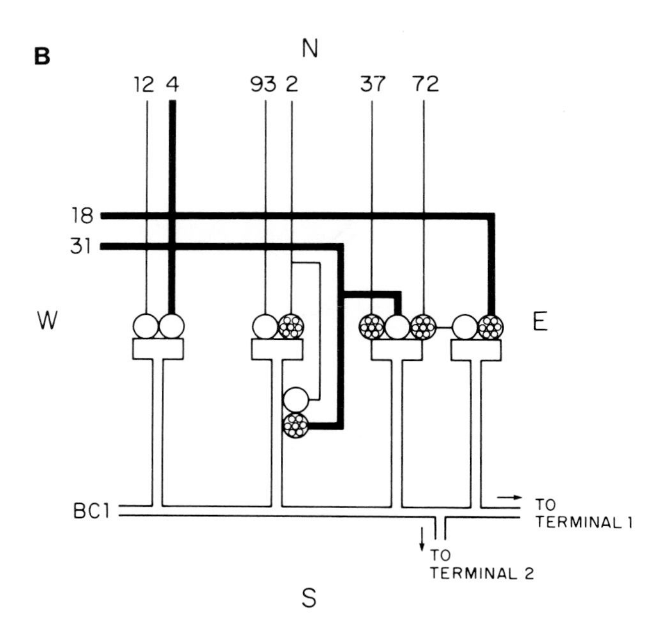

Fig. 4.8. Connection of the collaterals of bipolar cell 1 at terminal 1 (A) and terminal 2 (B). The approach directions of the large (thick lines) and the small (thin lines) horizontal cells are indicated. With the exception of one connection to efferent bipolar cell 14 at terminal 1, all connections involve horizontal cells.

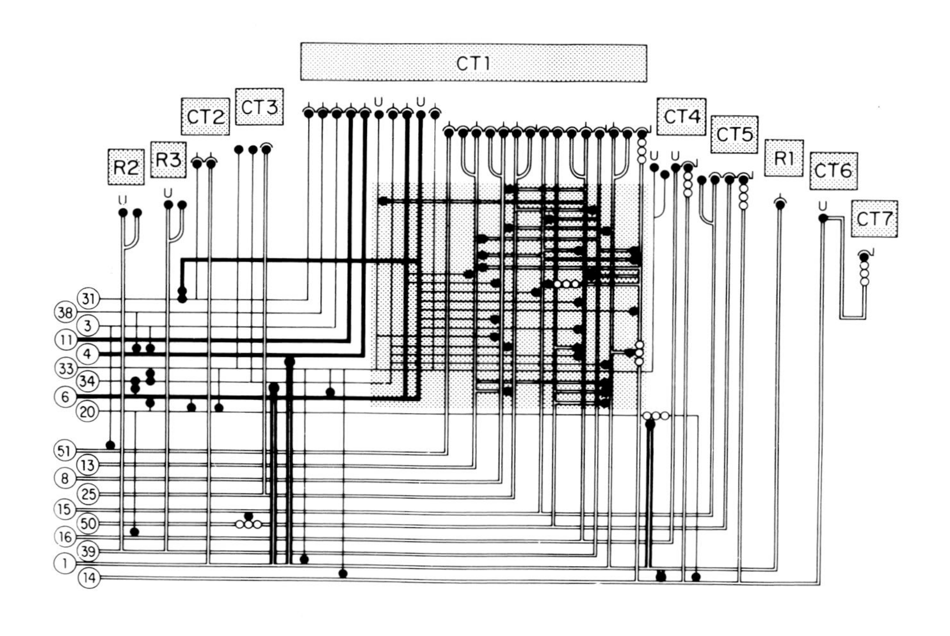

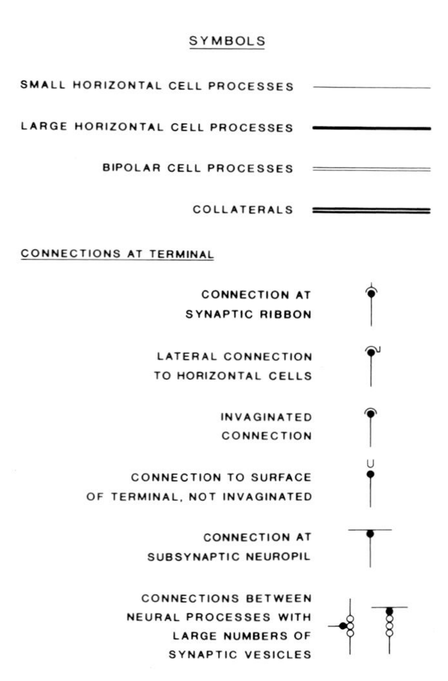

Fig. 4.9. Neural connections observed during analysis of terminal 1. Shaded area indicates the subsynaptic neuropil.

collaterals is that the collaterals as well as the connecting horizontal cell processes are considerably thicker than the processes passing through the subsynaptic neuropil to contact the terminal. The larger width favors a more rapid conduction in the collaterals than in the processes contacting the terminal. Furthermore, any influence exerted on bipolar cell 1 by horizontal cells in the common neuropil is transmitted at a level proximal to that at which the bipolar cell 1 process splits up into end branches to the terminals.

In Fig. 4.9 most neural connections observed during the analysis of terminal 1 are shown schematically. The subsynaptic neuropil connections at terminal 1 are indicated by the shaded area. The collaterals of bipolar cell 1 are emphasized by thick double lines.

Chapter 5: Brightness Contrast Enhancement

1. Interreceptor Connections and Connections between Cones and Bipolar Cells Connected to Rods

The first three-dimensional reconstruction of a part of the retina (Figs. 1.10–1.12) revealed interreceptor connections established by processes extending from cone terminals to rod terminals (Sjöstrand, 1958). At the rod terminals these processes contacted the terminal and neurons at the vitread pole, a location we now refer to as the subsynaptic neuropil (Fig. 5.1). One of these neurons was identified as a bipolar cell. Its end branch to the rod terminal was widened just outside the terminal. It then narrowed when passing through the opening of the invagination of the rod membrane. Within this invagination it widened again as it ended at the synaptic ribbon complex.

We will refer to the processes extending laterally from the cone terminals as cone processes. Some cone processes extended long distances, and all rod terminals were contacted by processes from several surrounding cones. Also, cones contacted each other through these processes (Sjöstrand, 1965). The analysis of the cone processes in the rabbit retina revealed that the connections of these processes differed at cone and at rod terminals. At cone terminals they contacted only the terminals but no neural processes in the subsynaptic neuropil, while at the rod terminals they were connected to the terminal as well as to neurons in the subsynaptic neuropil. There was therefore a striking difference in the pattern of connections of the cone processes at the two types of photoreceptors.

That the cone processes make contacts of functional significance is shown by the presence of synaptic vesicles in the processes at the area of contact. Synaptic vesicles are thus present at the ends of long processes even when there is a stretch along the process where vesicles are missing. In short processes synaptic vesicles are present along the entire process.

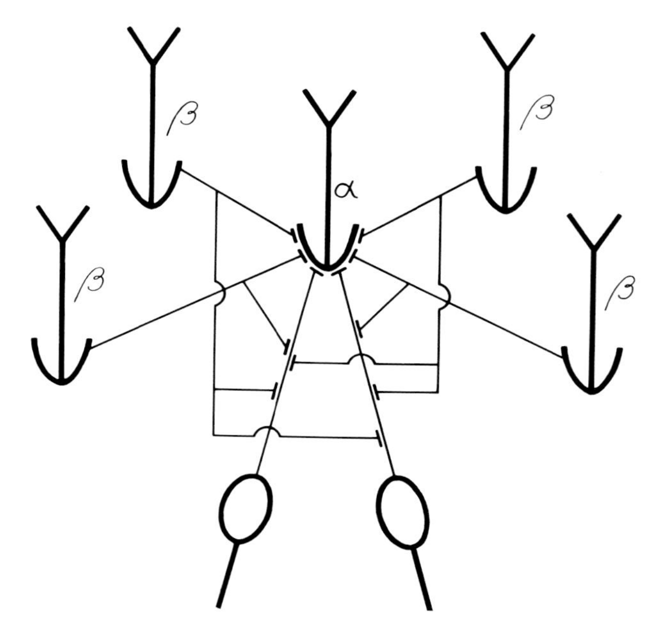

Fig. 5.1. The cone (β) connections to a rod terminal (α) and to bipolar cells connected to the rod, according to the first three-dimensional reconstruction of a photoreceptor terminal. From Sjöstrand (1958).

A detailed analysis of the connections of the cone processes at rod terminals in the rabbit retina by three-dimensional reconstruction of a number of rod terminals revealed that the connections of the cone processes in the subsynaptic neuropils were confined to bipolar cell end branches contacting the rod terminal and that there were no connections to horizontal cell end branches (Figs. 5.2 and 5.3). The widening of the bipolar cell end branches in the subsynaptic neuropil and the large size of the endings of the cone processes allowed the contact to involve large areas.

Synaptic vesicles were always present in the endings of the cone processes at the area of contact, and some small vesicles and small opaque granules were also present in the bipolar cell end branches in the subsynaptic neuropil (Fig. 5.4). The space separating the neural membranes at the area of contact was narrower than the space separating the neural membranes outside this area (Figs. 5.5 and 5.6).

Several observations reveal the presence of gap junctions at interreceptor contacts involving rods in nonmammalian retinas. In the toad retina,

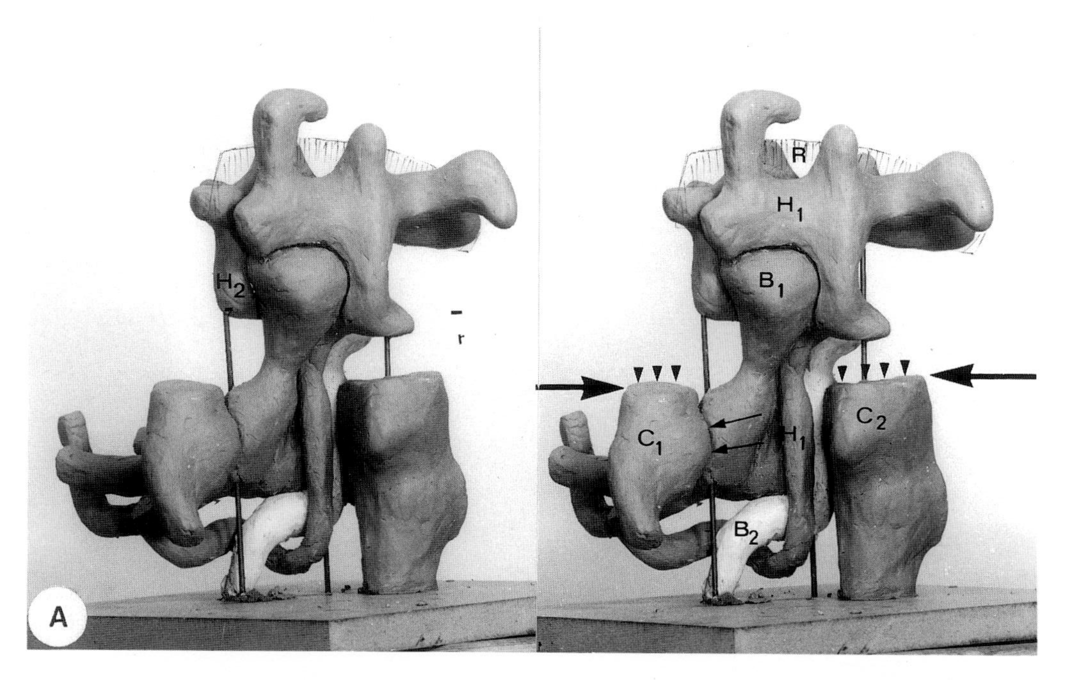

Fig. 5.2.A. Stereophotographs of three-dimensional reconstruction of synaptic ribbon complex and part of the subsynaptic neuropil at rod terminal Y in Fig. 1.20 (see also Fig. 2.16). In Fig. 5.2 and 5.3 the sclerad direction is upward. This view of the model reveals the extensive connections between bipolar cell B_1 and cone process C_1 (small arrows). The separation of cone process C_2 and the end process of horizontal cell H_1 is also shown. In Fig. 5.2 and 5.3 the large arrows indicate the location of the vitread surface of the rod terminal. Endings located above the plane indicated by these arrows are invaginated. Arrowheads indicate surfaces at cone processes contacting the surface of the terminal.

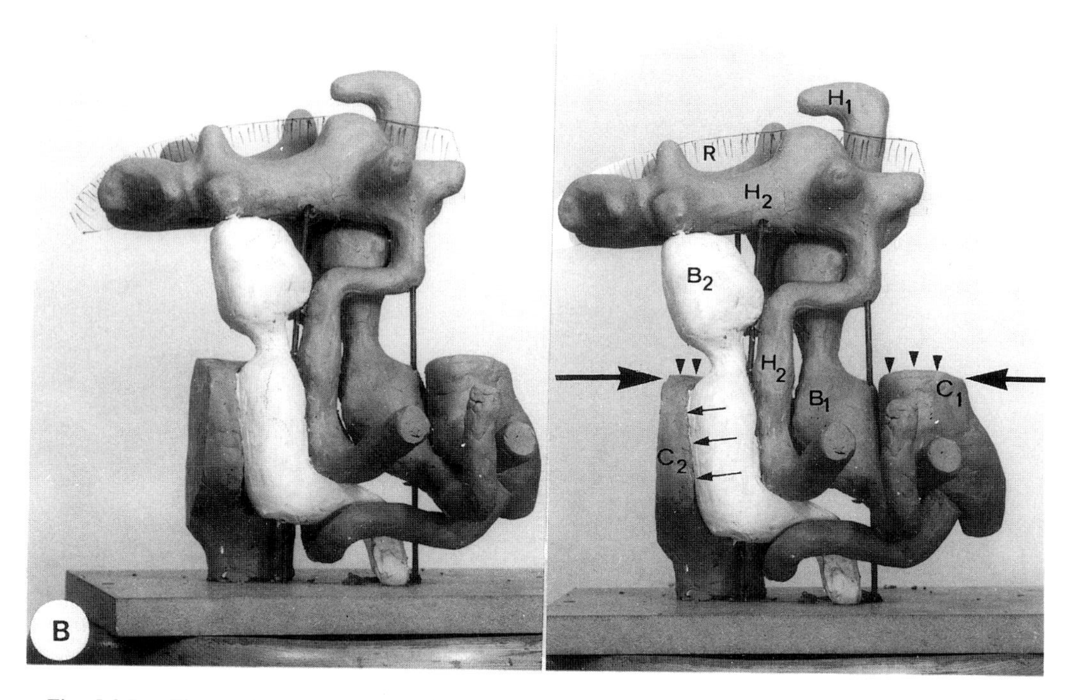

Fig. 5.2.B. The model has been turned almost 180° and now reveals the extensive connection between bipolar cell B_2 and cone process C_2 (small arrows).

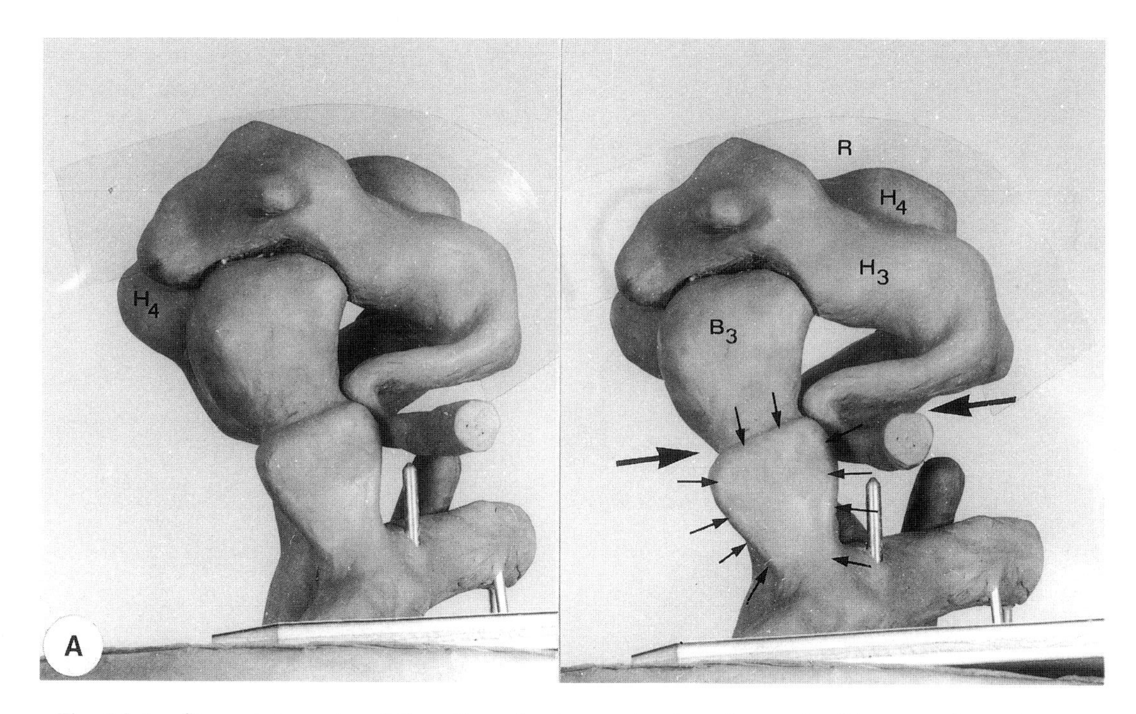

Fig. 5.3.A. Stereophotographs of three-dimensional reconstruction of synaptic ribbon complex and part of subsynaptic neuropil at rod terminal X in Fig. 1.20 (see also Fig. 2.17). Bipolar cell B_3 exposes a large surface, indicated by small arrows, connected to a cone process (not shown).

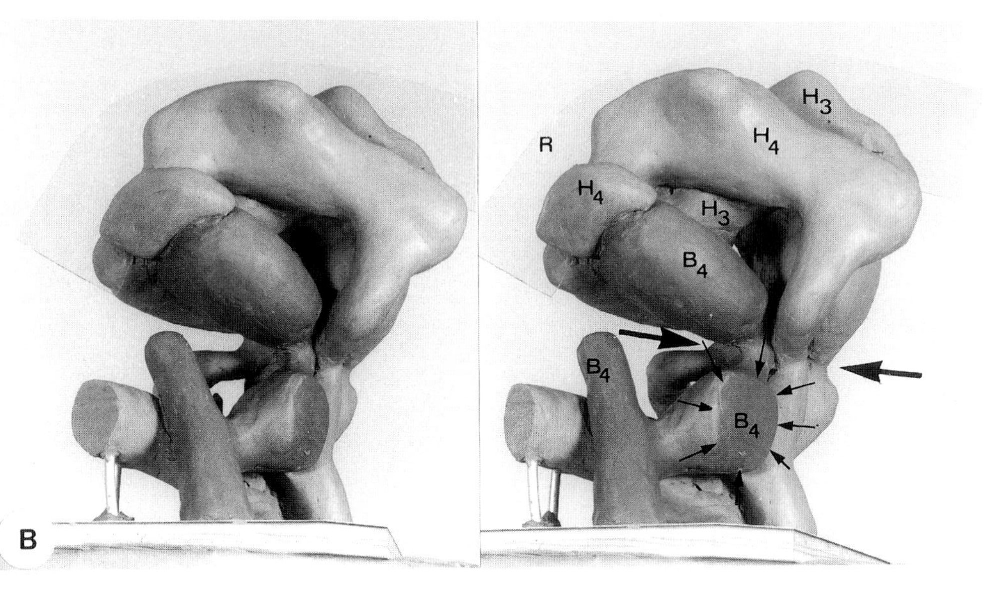

Fig. 5.3.B. A widened part of the bipolar cell B_4 ending exposes a large area, indicated by small arrows, connected to a second cone process (not shown). The cone connections involve cones located in opposite directions from the rod terminal.

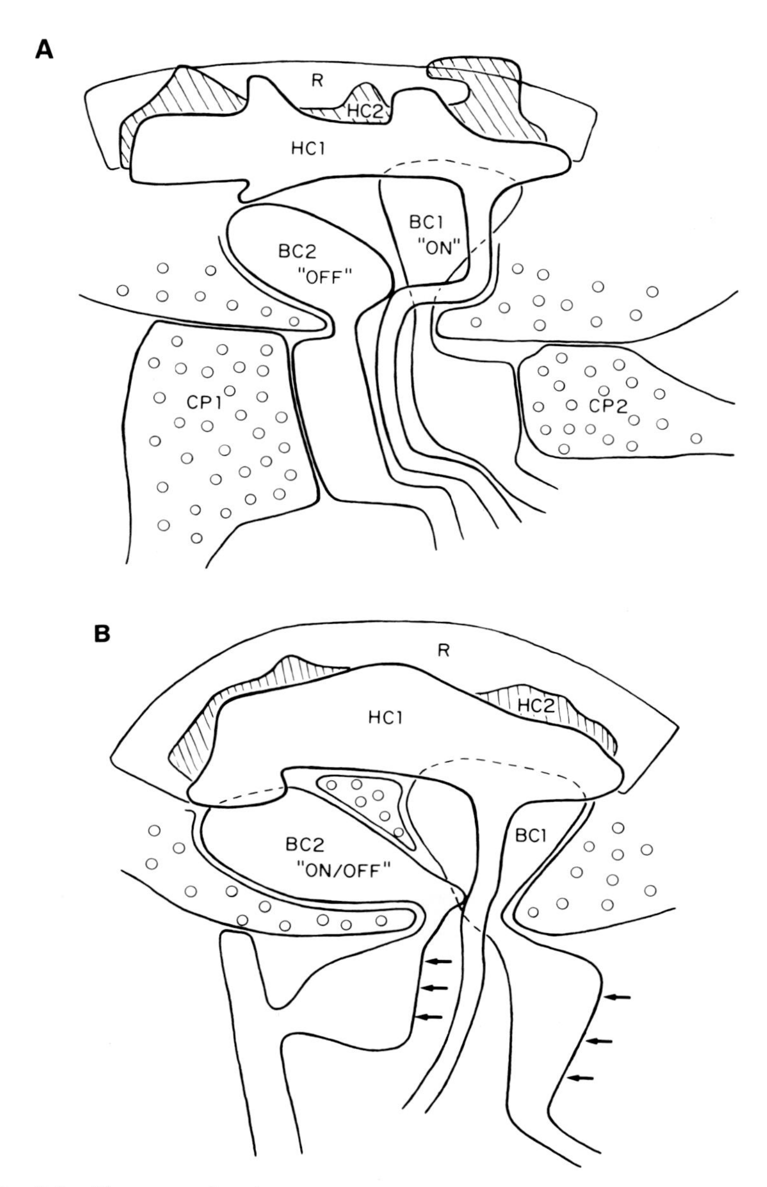

Fig. 5.4. The connections between cone processes and bipolar cell processes at the two terminals shown in Figs. 5.2 and 5.3. In A two cone processes, CP1 and CP2 contact bipolar cells BC2 and BC1 respectively as well as the surface of the terminal. In B the cone processes are omitted but the surfaces of bipolar cells BC1 and BC2 that contact cone processes are indicated by arrows. To simplify the drawing these surfaces have been drawn facing in the same direction. They really face in opposite directions in the model. The deduced functional characteristics of the bipolar cell endings are indicated, ''on'', ''off'' and ''on/off'' endings.

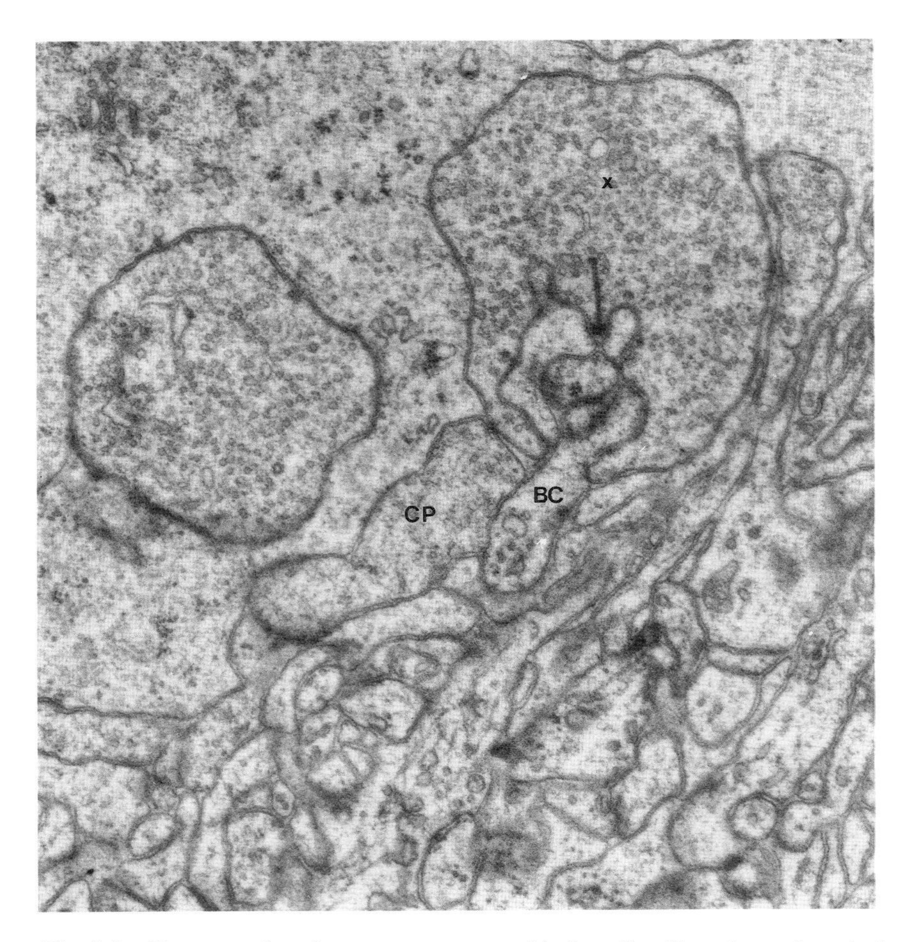

Fig. 5.5. The connection of one cone process to a bipolar cell end branch at rod terminal Y, the three-dimensional reconstruction of which is shown in Figs. 2.16 and 5.2. The cone process (CP) and the bipolar cell process (BC) are closely apposed. The connection between the cone process and the rod terminal is more extensive in the adjacent sections, with numerous synaptic vesicles present in the cone process ending. 30,000×.

red rods are connected by gap junctions at the level of the inner segment (Fain *et al.*, 1975), but no such connections were observed at the level of the terminals. The red rods were shown to interact functionally in this case.

Raviola and Gilula (1973) observed junctions resembling gap junctions at cone-to-cone contacts in freeze-fractured material. Whether these junctions are functionally electrical synapses remains to be shown.

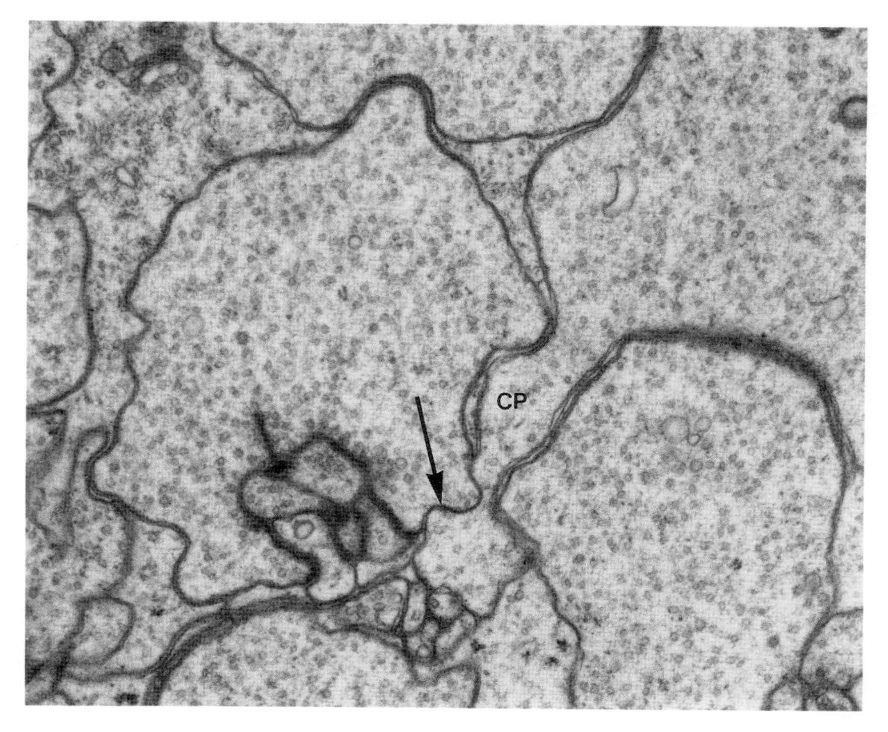

Fig. 5.6. Close association (arrow) of neural membranes of a cone process (CP) and a rod terminal in rabbit retina. In this specimen the synaptic vesicles in the horizontal cell endings at the synaptic ribbon complex are well preserved. Synaptic vesicles are clustered at the rod membrane facing the horizontal cell endings.

2. THE DOUBLE INPUT TO BIPOLAR CELLS CONNECTED TO RODS

The connections of the cone processes reveal that the bipolar cells connected to rods receive an input from both rods and cones. The latter input consists of one direct input to the dendritic end branches of the bipolar cells in the subsynaptic neuropil and one indirect input through the connections between the cone processes and the rod terminals.

The influence exerted by the cones on the rods will be of a facilitatory nature when both cones and rods are exposed to the same light stimulus. This influence may also affect the bipolar cell responses generated by the rods. However, this indirect influence is delayed by a synaptic delay and therefore does not affect the early stage of the bipolar cell response.

The responses of the bipolar cell to the direct input from the cones are not modulated by the horizontal cells. Instead, the bipolar cells are exposed to a continuous depolarizing pressure exerted by the cones with no opposing hyperpolarizing pressure modulating the bipolar cell responses. The bipolar cells would therefore be maintained in a depolarized state, creating extensive background noise in the absence of a balancing hyperpolarizing pressure.

This leads to the search for a type of neuron that can play a role similar to that of the horizontal cells in balancing the membrane potential of the bipolar cells connected to rods. With no such neuron present in the outer circuitry layer we have to look in the inner plexiform layer. In that layer the amacrine cells are structurally qualified for this function.

The amacrine cells show certain structural similarities to the horizontal cells. Their dendritic trees have a wide lateral spread corresponding to a large structural receptive field. They make triad connections involving bipolar cells and ganglion cells, although the connections are not invaginated. Instead there are plenty of Müller's cell processes that can insulate these connections. Synaptic vesicles present both in the bipolar cell axon and in the amacrine cell at the site of contact qualify the bipolar cell–amacrine cell connection as a reciprocal synapse similar to the reciprocal synapses between the horizontal cells and the photoreceptors.

To add further to the similarity there is a synaptic ribbon located in the bipolar cell axon at the site of the triad connection. There are therefore several structural features that point to the amacrine cells having a function similar to that of the horizontal cells.

The amacrine cells are also implicated from a functional point of view. As pointed out in Chapter 3, Section 24, Kuffler (1953) found that the surround can inhibit ganglion cell responses. Such inhibition has not been recorded at the bipolar cell level, where the surround instead exerts a facilitatory influence, for instance, on the depolarizing (on-) bipolar cells (Naka and Ohtsuka, 1975). The inhibition must therefore be mediated by neurons in the inner plexiform layer, where the amacrine cells are structurally the only neurons that could account for such lateral inhibition.

This inhibition can be explained by the amacrine cells releasing a hyperpolarizing transmitter, which is required for a modulation of the bipolar cell membrane potential. With the release of a hyperpolarizing transmitter the amacrine cells can modulate the bipolar cell and the ganglion cell membrane potentials at the triad connections between these neurons, and this modulation will be similar to that exerted by the horizontal cells in the outer circuitry layer. When the bipolar cell is depolarized it exerts an increased depolarizing pressure on the amacrine cell, and after a synaptic delay this cell responds by increasing its hyperpolarizing pressure on both the bipolar and ganglion cells. The depolarization of the bipolar cell is then partially reversed. Thus, the influence of amacrine cells should mimic that of the horizontal cells.

Such a modulating influence exerted by amacrine cells prevents the bipolar cells connected to rods from generating extensive background noise. It guarantees that only signals generated by discrete events at the level of the photoreceptors are transmitted to the ganglion cells.

The intrinsic bipolar cells that we assume are connected to amacrine cells may contribute importantly to a regulation of the hyperpolarizing pressure by a continuous depolarizing action on amacrine cells.

We can conceive of the function of the amacrine cells as being similar to that of the horizontal cells in that they serve as a gate that regulates the transmission of signals to the ganglion cells in such a way that background noise is eliminated.

The amacrine cells have been shown to respond with graded changes in membrane potential. A depolarization of a bipolar cell that is under the threshold for activating a ganglion cell therefore can still make an amacrine cell increase the hyperpolarizing pressure it maintains. One consequence of this is that an activation of a ganglion cell by the summation of subthreshold stimuli becomes unlikely because the amacrine cells increase their inhibitory influence on the ganglion cells when exposed to subthreshold stimuli. This eliminates background noise by eliminating a summation of subthreshold stimuli generated by randomized neural activity.

The depolarizing pressure exerted by the cones on on-bipolar cells connected to rods adds to that exerted by the rods. However, when both rods and cones are stimulated by light, the reduction of the depolarizing pressure exerted by the cones opposes the depolarization of the on-bipolar cells by the rods. Under this condition the cones act on the on-bipolar cells in an inhibitory way, raising the threshold for their activation.

At a very high illuminance, when the depolarizing pressure exerted on on-bipolar cells by both rods and cones approaches its lower limit, we expect that the inhibitory action of the cones on the bipolar cells connected to rods may be strong enough to block the transmission of signals by the on-bipolar cells without blocking the response of the rods to light stimuli, which instead is enhanced. This means that the horizontal cells may still receive an input from rods at a range of illuminance above that at which the responses of bipolar cells connected to rods are blocked by the cones. The horizontal cells therefore continue to contribute to the modulation of the responses of bipolar cells connected to cones on the basis of the input from both rods and cones.

It is important to realize that there is no corresponding inhibitory influence exerted by cones on bipolar cells connected to cones because cone processes are not connected to bipolar cells at the cone terminal.

Rod vision is consequently associated with a higher threshold than cone vision. A change in this situation at low illuminance requires that the inhibitory influence of cones be eliminated. Such a change in the cone influence will occur as a consequence of the cones approaching maximal depolarization at a low illuminance. The depolarizing transmitter is then released at a maximum level, keeping the receptor sites in the postsynaptic bipolar cell membrane saturated. According to the Michaelis–Menten relationship that we applied to transmitter–receptor site interaction (Chapter 3, Section 9), the modulated depolarizing pressure exerted by the cones on the bipolar cells connected to rods will change to a more or less constant pressure at a low illuminance. As a consequence, the cones will not raise the threshold of these bipolar cells. Instead, the constant depolarizing pressure will lower the threshold for rod vision by opposing the strong hyperpolarizing pressure exerted by the horizontal cells on the bipolar cells.

If the intensity of the light stimulating rods increases while the intensity of the light stimulating adjacent cones decreases, the cones will exert a facilitatory influence on the bipolar cells connected to rods. Such a combination of light stimuli will occur at boundaries between fields of different illuminance in the images projected onto the retina.

In the case of off-bipolar cells, the cones will facilitate the off-response when the intensity of the light stimulating both rods and cones is reduced. When the light intensity is increased, the cones will facilitate the hyperpolarization of off-bipolar cells connected to rods.

3. A Mechanism for Enhancement of Spatial Brightness Contrast: The Mach Bands

The image that we perceive through our vision is modified considerably from the pattern of illumination that gives rise to the perceived image. Certain pieces of information in the image are enhanced while others are eliminated. Thus, "the eye itself sacrifices accuracy about information of little consequence, such as the absolute levels of illumination, in order to enhance features that are more significant such as contours and edges" (Ratliff, 1965). The enhancement of boundaries facilitates the observation of objects at various conditions of illumination and against different types of background.

This enhancement of the contrast has been thought to be accomplished by a peripheral neural mechanism in the retina involving lateral inhibition or by a central mechanism in the brain. A combination of retinal and central factors has also been considered.

In 1865 Ernst Mach discovered a contrast phenomenon involving the appearance of dark and bright bands in the image of two fields of different brightness that are separated by a zone within which the brightness changes gradually from bright to dark. A bright band is located at the edge of the bright field and a dark band at the edge of the dark field. In this way the location of the boundaries of the two fields is more clearly perceived than it would be without this effect.

These bands, called the Mach bands, are more pronounced at steeper transition gradients. In the case of an abrupt change in the brightness the phenomenon leads to a perception of the bright field being brighter and the dark field being darker at the boundary separating the two fields, an enhancement of contrast at the boundary. This effect is illustrated in Fig. 1.3.

Mach analyzed this phenomenon in detail and proposed that it is due to a retinal mechanism that he assumed to involve lateral inhibition in the neural network of the retina. Mach believed that the significance of this distortion was to enhance the appearance of contours and borders. The possibility that the contrast effect was caused by successive contrast associated with the eye movements was ruled out by the observation that the Mach bands do not fade when the image is stabilized on the retina (Riggs *et al.*, 1961).

Hartline (1949) observed in the *Limulus* retina that the responses to a light stimulus of one ommatidium could be inhibited by the stimulation of an adjacent ommatidium. This observation showed that a mechanism for lateral inhibition exists in the *Limulus* retina. Hartline and Ratliff (1957) analyzed this effect further in experiments that mimicked the response to the projection of a boundary between two fields of different illuminance onto the *Limulus* retina and recorded responses corresponding to a boundary effect. The center–surround antagonism discovered by Kuffler (1952, 1953) in the vertebrate retina has been assumed to account for lateral inhibition responsible for enhancement of spatial brightness contrast.

When Sjöstrand (1958) discovered the interreceptor connections, he proposed that they are involved in lateral inhibition and account for the Mach band phenomenon in the vertebrate retina. At that time the mode of transmission between the neurons in the retina was not known, so no mechanism for the proposed lateral inhibition could be proposed. It was, however, considered to be advantageous that the lateral inhibition occurs at the level of the photoreceptors, because the inhibition could then be

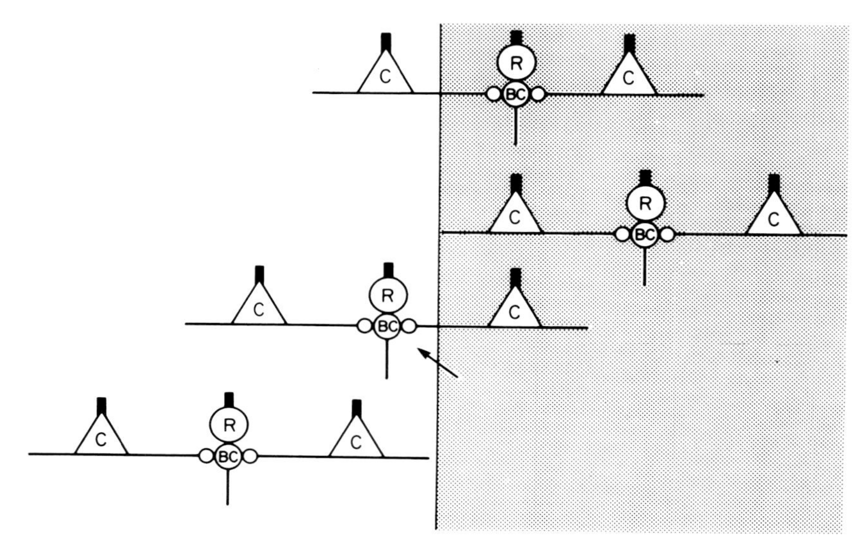

Fig. 5.7. The boundary effect, as explained by connections between the cones and the bipolar cells connected to rods.

confined to a small number of photoreceptors, a prerequisite for brightness contrast enhancement at a high resolution.

The observation that the cone processes contact the bipolar cell end branches in the rod subsynaptic neuropil and the deduction of the neural interactions that distinguish hyperpolarizing and depolarizing bipolar cells now make it possible to discuss a mechanism responsible for spatial brightness contrast enhancement.

Let us start by pointing out that, when maintained at a constant level or increased, the depolarizing pressure exerted by cones on bipolar cells connected to rods facilitates the bipolar cell responses. As discussed earlier, when both rods and cones are stimulated by light, the reduction of this depolarizing pressure acts in an inhibitory way by raising the threshold for bipolar cells connected to rods. The cones therefore inhibit these bipolar cells. The conditions are then fulfilled for lateral inhibition within the outer circuitry layer.

We will now discuss how this lateral inhibition can enhance spatial brightness contrast by considering a situation in which the image of two fields of different illuminance separated by a sharp boundary are projected onto the retina. We assume that the illuminance of both fields exceeds the preceding illuminance over the retina (Fig. 5.7). We can distinguish between the situation at the moment the image of the two fields is projected onto the retina and the situation when the two fields are viewed steadily. In the latter situation we can distinguish between the areas covered by the two fields away from the boundary and the area at the boundary.

When the image is projected onto the retina the on-bipolar cells become depolarized and the amplitude of the depolarization of on-bipolar cells connected to rods is reduced by the simultaneous reduction of the depolarizing pressure exerted by the cones.

At the boundary, the on-bipolar cells connected to rods located on the bright side are connected to cones located on both sides. Of these cones, those on the dark side exert a stronger depolarizing pressure on the on-bipolar cells than those on the bright side. The amplitude of the depolarization of the bipolar cells connected to rods on the bright side of the boundary is therefore larger than the amplitude of depolarization of the bipolar cells connected to rods located away from the boundary in the bright field, the bipolar cells of which are connected only to cones located within the bright field. The bright field consequently appears brighter at the boundary than away from the boundary.

On-bipolar cells connected to rods located at the boundary on the dark side are also connected to cones located on both sides of the boundary. The cones on the bright side exert a weaker depolarizing pressure on the bipolar cells than those on the dark side. As a consequence, the amplitude of the depolarization of these bipolar cells is smaller than that of on-bipolar cells connected to rods located away from the boundary in the dark field. This field therefore appears darker at the boundary than away from the boundary.

After the image has been projected and is viewed steadily, the perception of the image will depend upon the activation of on- and off-responses of bipolar cells connected to rods located at the boundary, where the intensity of the light stimulating the photoreceptors varies as a consequence of the involuntary eye movements.

Away from the boundary the image of the two fields is transmitted by L-bipolar cells. The conclusions that apply to the depolarizing bipolar cells when the image is first projected onto the retina apply to the L-bipolar cells as well. These bipolar cells will therefore also contribute to contrast enhancement.

According to this discussion, spatial brightness contrast enhancement can be explained on the basis of the connections between cones and bipolar cells connected to rods. This contrast enhancement is, however, confined to rod vision because the cones make no connections to bipolar cells connected to cones.

It has been difficult to establish whether the Mach band effect exists in the perception of colors in the absence of a difference in illuminance, so the existence of a mechanism for spatial contrast enhancement in color vision has therefore been questioned. Structurally, there are no neural connections that could account for the required neural interaction. The connections between cone terminals are of a facilitatory nature and can contribute to an enhancement of color contrast by increasing the saturation of the perceived colors. This requires that cones with the same spectral sensitivity be connected, which has been demonstrated by electrophysiological recordings from cones (Baylor, 1974).

Within a field that is uniformly illuminated with red light adjacent to a field illuminated with blue light, the facilitatory interaction between the red-sensitive cones will enhance the saturation of the perceived red color within the red field. At the boundary of this field the enhancement is weakened by the interaction with red-sensitive cones in the blue field that exert a depolarizing influence on red-sensitive cones in the red field. We therefore expect that the color contrast at the boundary should be reduced.

However, the perception of color difference at the two sides of a boundary seems to be reinforced by color enhancement due to successive contrast as the boundary oscillates as a result of involuntary eye movements. This enhancement can lead to a disturbing exaggeration of the color contrast at the boundary between, for instance, a red and a blue field of similar illuminance. In this case successive contrast alternately enhances the saturation of the red and the blue colors at the border in such a way that the color at the border flickers between a saturated red and a saturated blue.

This color flicker effect is pronounced and easy to observe as a disturbing phenomenon when a sharp boundary separates two fields with the proper solid colors. It seems possible, however, that the effect occurs less markedly at any boundary that separates two colors that activate two types of cones with the proper spectral sensitivities and that the effect then contributes to an enhancement of the border without any noticeable flicker.

We can conclude that vision involves a mechanism for enhancement of boundaries and of contours, that is, a mechanism for spatial brightness contrast enhancement. Contrast enhancement is associated with differences in illuminance; a corresponding enhancement of chromaticity gradients is uncertain.

The retina contains an elaborate system of cone processes extending laterally from the cone terminals to rod terminals and other cone terminals. In the subsynaptic neuropil at rods these processes contact bipolar cell dendritic endings but no horizontal cell endings. No corresponding bipolar cell connections exist at cone terminals. According to our deductions the connections of the cone processes to the bipolar cells at rod terminals can explain the Mach band phenomenon and can constitute a basis for spatial brightness contrast enhancement.

For this mechanism to work both rods and cones must function together. The supposed association of rod and cone vision with different intensities of ambient illuminance should preclude the interaction between the two types of photoreceptors required for the proposed mechanism for contrast enhancement. However, the confining of rod and cone vision to different ranges of illuminance is not justifiable. Aguilar and Stiles (1954) thus observed that rods can adapt over a wide range of ambient illumination, 1000 trolands. This means that the rods are active in daylight vision. In fact, pure cone vision may occur only at light intensities that in the long run are intolerable (Weale, 1961). Under normal conditions of illumination, both rods and cones must be active over a wide range of ambient illumination, i.e., vision is mesopic.

The Purkinje shift has been interpreted as indicating a switch from rod vision to cone vision when the illuminance exceeds a rather low intensity. This shift is based on measurements of thresholds, which show nothing more than that at a low illuminance the threshold for rod vision is low and the threshold for cone vision is high and that the opposite applies at high illuminance.

If the concept that rod vision is confined to low illuminance were correct, about 100 million photoreceptors, that is, about 95% of all photoreceptors in the human eye, would not function under most conditions of illuminance. Such a waste of neural matter is difficult to reconcile with the economy characterizing the designs of Nature, and the concept appears silly.

At a low illuminance we expect that the spatial brightness contrast enhancing mechanism would not function because the modulating influence of the cones on the bipolar cells connected to rods is reduced or is eliminated when the cones approach maximal depolarization, as discussed in section 2 of this chapter. This agrees with the observation that the Mach bands are not observed at a low illuminance. The elimination of the spatial brightness contrast enhancing mechanism therefore contributes to the deterioration of the quality of the perceived image under these conditions.

The efficiency of the contrast enhancing mechanism is determined by the minimum slope of the gradient between two fields of different illuminance required to produce the Mach bands. Within an intermediate range of illuminance this slope is constant while at high illuminance an increasing slope is required. The contrast enhancing mechanism thus becomes ever less efficient at higher illuminance. This phenomenon is explained by the reduction of the contribution of the rods to vision at very high illuminance, where we can expect that the inhibition of rod vision by cones eventually blocks rod vision.

Spatial brightness contrast is enhanced particularly strongly in the human fovea, and to explain this enhancement on the same basis proposed for the rabbit and the guinea pig retinas one must assume that cone processes in the fovea contact bipolar cells that are connected to cones.

The fovea is designed for optimal quality of vision which requires maximum visual acuity, spatial brightness contrast enhancement, and contrast enhancement by color discrimination. As is discussed in Chapter 7, Section 7, the association of spatial brightness contrast enhancement with two types of photoreceptors leads to an enormous saving of neural matter because only a limited population of photoreceptors must be equipped with processes extending laterally to establish contacts to bipolar cells at the terminals of other photoreceptors. This saving of neural matter is achieved at the expense of the amplitude of the distortion of the contrast at boundaries because the larger the population of photoreceptors that can contribute to lateral inhibition, the more strongly the spatial brightness contrast can be enhanced. We would expect that maximum enhancement occurs when all photoreceptors are involved in lateral inhibition. For this to occur, all cones in the fovea must extend processes contacting bipolar cell end branches to cones. The enhancement of spatial brightness contrast will then require an expenditure of neural matter considerably larger than that of the extrafoveal retina.

A pure cone population creates optimal conditions for spatial color contrast at a high resolution. However, the circuit for spatial brightness contrast should be kept separate from that for color contrast to prevent distortions of the perceived colors. The color contrast is dependent primarily on the saturation of the colors, and any interference with the mechanism for saturation enhancement must be avoided.

There is another reason for separating these circuits. The lateral inhibition lowers the upper limit of the range of illuminance over which vision functions, and any such reduction is prevented if the circuit for spatial brightness contrast enhancement is separated from the circuit for color discrimination because then lateral inhibition will not involve bipolar cells transmitting color-coded signals.

To satisfy these requirements the information regarding brightness should be transmitted by bipolar cell channels different from those that transmit information regarding colors. This means that each cone should be connected to one bipolar cell representing the "color channel" or chromatic channel and to a second bipolar cell representing an achromatic channel.

As a consequence of cone processes extending only a short distance to bipolar cells connected to adjacent cones, the boundary effect is confined to narrow zones that correspond to a high "resolution" of the boundary effect. The close packing of the cones also contributes to a further enhancement of the spatial brightness contrast.

We have only crude information regarding the morphology of the terminals of the photoreceptors in the fovea. Cohen (1972) referred to the photoreceptor terminals in the human fovea as pedicles with short processes that contact all surrounding pedicles. The fact that these processes are not confined to a single plane may indicate that they also enter the subsynaptic neuropil to establish inhibitory connections to bipolar cells. Ramon y Cajal (1892) described the fovea of the greenfinch (*Carduelis chloris*) retina in Golgi-impregnated material. He described the synaptic terminals within the fovea as having a conical or an ellipsoid shape without any basal filaments (cone processes) or with some short "rudimentary" filaments. There are thus definite differences in the size and the shape of the terminals between cones in the fovea and cones outside the fovea. These descriptions may conform with the design suggested above that includes facilitatory connections between cones as well as inhibitory connections from cones to bipolar cell end branches to cones.

4. Electrophysiological Confirmation of Interreceptor Connections

The interpretation that the observed interreceptor connections revealed functional interactions between photoreceptors was originally not accepted by the electrophysiologists because no electrophysiological observations had revealed such interactions. The electrophysiologists were not impressed when told that they had not studied the behavior of the photoreceptors exposed to light stimuli sufficiently restricted spatially to correspond to the range over which the interreceptor connections extend. The attitude of the electrophysiologists was understandable. The resolution at which they could analyze retinal functions at that time (ganglion cell response) was not good enough for analysis of this problem.

In 1971, thirteen years after the discovery of the interreceptor connections, Baylor and co-workers confirmed the presence of such connections in the turtle retina. These experiments involved intracellular recordings from single cones. For optical reasons it was not possible to stimulate the impaled cone selectively. The stimulus therefore also included a population of cones surrounding the impaled cone. For this reason, as pointed out by Baylor *et al*, the recorded responses were composite responses and consisted of one component due to the light stimulating the impaled cone and a second component initiated within a small surrounding area of the retina.

The interreceptor interaction was shown by the increase in the response amplitude when the circle of light falling on the retina increased in radius from 2 to 8 μm. This effect was interpreted as evidence for facilitation exerted on the impaled cones by surrounding cones.

An interreceptor interaction was also verified in experiments in which electrodes were inserted into two neighboring cones and current was run through either cell. An inward current caused hyperpolarization whereas an outward current caused depolarization of the neighboring cone. These responses agreed with the response evoked by light. In one experiment, however, the depolarizing current through one cone, hyperpolarized the other cone.

These experiments led to the conclusion that the interaction between the cones involves facilitation because in most cases the change in membrane potential in both neighboring cones was of the same sign. According to Baylor *et al.* (1971), the experiments did not allow any conclusion as to whether the interreceptor connections improve or impair spatial contrast.

Later, Baylor (1974) showed that cones were functionally connected in a systematic way, and that the connections were confined to cones with the same spectral sensitivity. The functional significance of these cone-to-cone connections still appeared obscure. However, the cones contribute to contrast enhancement by color discrimination, and this discrimination becomes better as the perceived colors become more saturated. The interreceptor connections can thus still be considered to be involved in contrast enhancement.

Inhibition assumed to involve cones inhibiting rods and vice versa has been demonstrated by Rodieck and Rushton (1976) in recordings from single ganglion cells in the cat retina. These experiments did not allow any conclusions to be drawn regarding the level in the retinal circuitry at which the mutual two-way inhibition was exerted.

Chapter 6: Structural Basis for Some Other Neural Interactions

1. Directional Selectivity: Physiological Observations

A directionally selective unit responds to an image moving over the retina in one particular direction, the preferred direction, while it does not respond to an image moving in the opposite direction, the null direction. To qualify, the directionally selective unit must show true selectivity for direction in all parts of the receptive field and must be independent of the image contrast. Thus, it responds in an identical way whether the image is dark or bright.

Directionally selective units were first observed in intracellular recordings from neurons in the cerebral cortex of cats (Hubel, 1959; Hubel and Wiesel, 1959, 1968) and in the optic tectum of frogs and pigeons (Lettvin *et al.*, 1959; Maturana and Frenk, 1963). Barlow and Hill (1963) showed that coding for direction of movement can be carried out by the rabbit retina by discovering directionally selective ganglion cell responses in this retina. That the responses were representative of a specific type of neural interaction within the retina and not due to asymmetry in the organization of the receptive field or to artefacts was also shown by Barlow and Hill (1963).

Barlow, Hill, and Levick (1964) showed that the directionally selective responses of ganglion cells could be evoked by movements of an image over a distance considerable shorter than the diameter of the center of a ganglion cell's receptive field. They therefore concluded that the basic units of directional selectivity are a set of subunits within the receptive field and that these units precede the ganglion cell, this latter cell representing a common pathway. The minimum distance required for the movement of the image to generate a response is less than 50 μm across the retina (Barlow and Levick, 1965). These observations led to the suggestion that the directionally selective unit could be a bipolar cell.

Directionally selective responses are evoked by an image moving over the center of the receptive field of a ganglion cell. However, two static stimuli separated in time can evoke the same responses provided that the spatial separation of the stimuli is small enough (Barlow and Levick, 1965). When the separation exceeds 220 μm the difference between a sequence of the two stimuli in the preferred and in the null direction becomes less evident. This observation shows that the directional selectivity responses are evoked by sequential stimulation of areas within the receptive field of the ganglion cell. The small spatial separation of static stimuli required to evoke the response conforms with the concept that the real unit for this type of response is the bipolar cell and not a population of bipolar cells.

The directionally selective ganglion cells are characterized by on/off-responses; units generating on-responses are rare (Barlow and Levick, 1965).

The directional selectivity was established to be due to inhibition generated by the movement of the image in the null direction although indications were also found for the participation of facilitation enhancing the response of the ganglion cells to movement in the preferred direction. The latter component was found to be involved in the responses of the directionally selective units in the ground squirrel retina (Michael, 1968).

Wyatt and Daw (1975) clearly revealed the involvement of lateral inhibition. These authors recorded extracellularly from the ganglion cell layer, and their recordings therefore involved a population of ganglion cells. By flashing a spot of light on one point in the receptive field and by moving a second spot along several parallel tracks across the field in the null direction, they found that inhibition was induced from areas in the receptive field lying predominantly in one direction from the flashing spot of light. A light spot that was moved in the opposite direction did not provoke any inhibition.

Because the recordings in this study were extracellular, they are difficult to evaluate quantitatively. The area in which inhibition occurs, however, is considerably smaller than the center of the receptive field of a ganglion cell, and it extends out into the surround when the stimulated stationary spot is close to the boundary of the receptive field center at its null side.

The electrophysiological recordings reveal that inhibition is the basis for the directionally selective response and that facilitation may be involved in the reaction for a movement of an image in the preferred direction. We will now consider bipolar cell 1 as a possible part of the circuitry for this type of response.

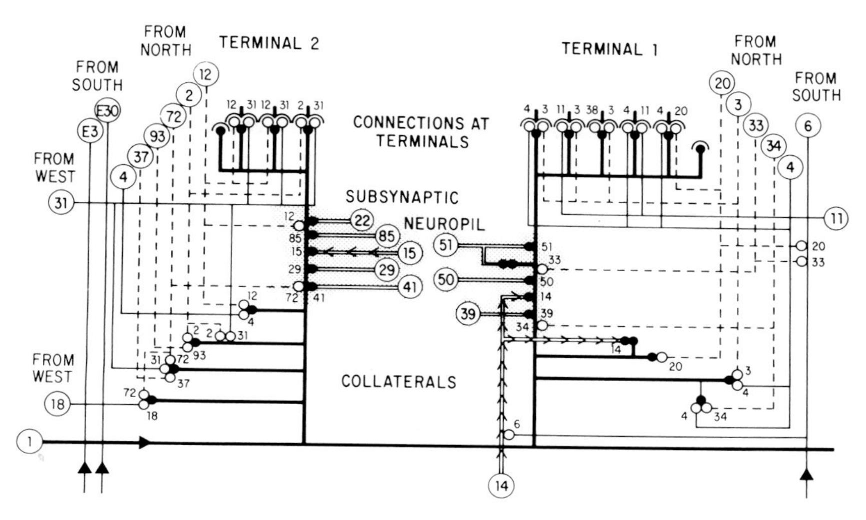

Fig. 6.1. Simplified representation of the various neural connections of bipolar cell 1 at terminals 1 and 2, in the subsynaptic neuropils of these terminals (shaded) and in the common neuropil. Thick line, bipolar cell 1; thin lines, large horizontal cells; broken lines, small horizontal cells; double lines, bipolar cells; arrowed double lines, efferent bipolar cells.

2. Bipolar Cell 1: A Special Type of Bipolar Cell

The most rational way to classify a neuron is by its function. Since the connections of a neuron in a neural circuit determine its function, we come close to a functional classification of a neuron by classifying it on the basis of its neural connections. With a sufficient knowledge of the function of the circuit, one can translate the pattern of connections into one or several functions of the neuron.

Bipolar cell 1 offers an example of this approach. We have considerable information regarding the neural connections of this bipolar cell at terminal 1 and at terminal 2 as well as in the common neuropil. Figure 6.1 summarizes the observations made so far. The connections of bipolar cell 1 make this neuron stand out as a bipolar cell with a special function differing from that of the other bipolar cells. These connections suggest one particular function for this neuron even though all its connections have not been revealed and even though the functional significance of all observed connections cannot yet be deduced. Let us first summarize the known neural connections of bipolar cell 1.

(1) Bipolar cell 1 had synaptic connections with many neurons at three different levels, the terminals, the subsynaptic neuropils,

and the common neuropil. It was the only bipolar cell that contacted other neurons in the common neuropil through special end branches, the collaterals.

(2) Most contacts with horizontal cells at the terminal and at the collaterals involved one small horizontal cell that was paired with a large horizontal cell. In the subsynaptic neuropils, on the other hand, the connections involved primarily other bipolar cells mostly in association with one horizontal cell.

(3) All endings of small horizontal cells contacted by bipolar cell 1 were the endings of branches that extended from main horizontal cell processes approaching the terminals from the north.

(4) None of these small horizontal cell processes contacted any photoreceptor located south of the terminals.

(5) No large or small horizontal cell processes approaching from south contacted bipolar cell 1.

(6) Most small horizontal cell processes that contributed end branches to terminal 1 and 2 either were contacted by large horizontal cell processes that approached the terminals from south or contacted such processes with end branches (Fig. 6.2).

(7) Bipolar cell 1 contacted the core bipolar cells at both terminals 1 and 2. These contacts were analyzed in detail in the subsynaptic neuropil of terminal 2. There were four contacts between these neurons, and at all four contacts there was an accumulation of synaptic vesicles in the processes of bipolar cell 1, while there were no synaptic vesicles in the contacting branches of the core bipolar cell, in contrast to the contacts made to other neurons by the core bipolar cell. In all these other contacts the core bipolar cell processes contained synaptic vesicles at the site of contact. The situation was similar at terminal 1, where the core bipolar cell also contacted bipolar cell 1.

(8) In the subsynaptic neuropils of both terminals 1 and 2, efferent bipolar cells contacted bipolar cell 1 and at both terminals synaptic vesicles were present in the contacting end branches of the efferent bipolar cells. Bipolar cell 1 therefore may be under feedback control from the inner plexiform layer.

(9) Bipolar cell 1 was also contacted by intrinsic bipolar cells other than the efferent bipolar cells. The connections in the subsynaptic neuropil were of the same type at both terminals 1 and 2. They therefore represent a part of a systematic pattern of connections. A more complete analysis of the circuitry including the inner plexiform layer is technically possible and would reveal the

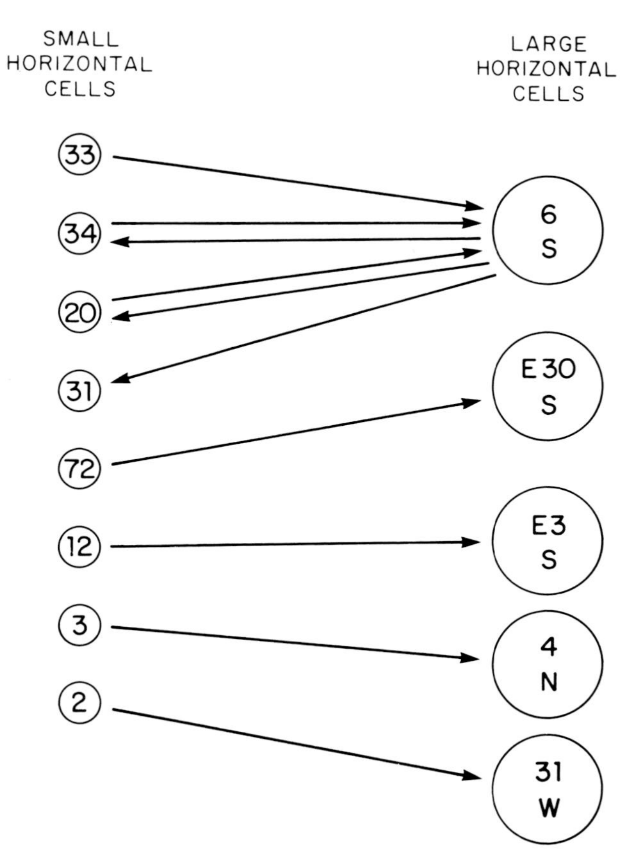

Fig. 6.2. Connections between small and large horizontal cells within the reconstructed region. These connections are established either by end branches from small horizontal cells or by special branches from the large horizontal cells as indicated by arrows. Small horizontal cells 2, 3, 12, and 72 belong to the reconstruction of terminal 2, as do large horizontal cell processes E30, E3, and 31.

overall pattern of neural connections of the bipolar cell 1 type of neuron.

(10) The two cone terminals and the rod terminal connected by bipolar cell 1 were lined up in an east–west orientation, that is, perpendicular to the approach direction of the small horizontal cells connected to bipolar cell 1. Three photoreceptor terminals, however, are too few to prove that this orientation was representative of the overall orientation of the dendritic end branches of bipolar cell 1.

We can now distinguish between connections of bipolar cell 1 and horizontal cells at synaptic ribbon complexes and in the common neuropil and connections between bipolar cell 1 and other bipolar cells in the subsynaptic neuropil. The latter, rather complex, connections involving several different types of bipolar cells, are most astonishing because the occurrence of elaborate connections between bipolar cells in the outer plexiform layer does not conform with the generally accepted function of this layer. These connections show that bipolar cells can interact laterally in this layer and that it is highly likely that they experience feedback control by neurons in the inner plexiform layer.

The lateral interaction between bipolar cells introduces a new significance to the wide spread of the dendritic tree of most bipolar cells because it shows that this spread can fulfill a function quite different from that of summation of the input from photoreceptors. The large dendritic trees of the bipolar cells, which allow extensive summation, has been interpreted as meaning that a bipolar cell behaves "as though there existed a broad photoreceptor of this diameter" (Rodieck, 1973). This appears to be an unjustifiable oversimplification based on information too crude to allow any conclusions regarding the factors involved in shaping the responses of the bipolar cells.

A large number of bipolar cells contact each cone terminal and bipolar cells can contact both cones and rods. There are usually more than two bipolar cells contacting each rod terminal. These observations mean that the dendritic trees of bipolar cells must overlap extensively. There are no reasons to assume that this situation is unique to the rabbit retina. The conclusion based on the analysis of Golgi-stained neurons that the overlap of bipolar cells is small, 10%, (Boycott and Dowling, 1969) therefore is not supported by the detailed analysis of neural connections that the three-dimensional reconstructions make possible. The claim of Boycott and Dowling can be explained by the analysis involving incompletely stained bipolar cells. This is a major deficiency of the Golgi staining technique and it further limits the conclusions that can be drawn on the basis of studies of such material.

The observation that bipolar cells can be connected synaptically to both rods and cones agrees with that of Polyak (1941), which was confirmed by Stell (1965). These connections must have an important function because they allow neural circuits to function at low ambient illuminance.

We conclude that bipolar cell 1 is unique among the eighteen bipolar cells that were connected to terminals 1 and 2. The connections of bipolar cell 1 are very instructive because they reveal that bipolar cells are synaptically connected in a way considerable more complex than we would have expected if their only function were to sum and transmit signals from photoreceptors.

3. A Circuit for Directionally Selective Bipolar Cell Responses?

The special neural connections at bipolar cell 1 suggested that this type of bipolar cell could be involved in directionally selective responses. The fact that bipolar cell 1 is an on/off-unit qualifies it for this function. The asymmetry of the connections to small horizontal cells and the fact that these cells can transmit information regarding the illumination in only one part of the receptive field of bipolar cell 1 also seems to qualify this type of bipolar cell for this function.

The collaterals that we interpreted as allowing the horizontal cells to exert a strong influence on bipolar cell 1 seem to satisfy the requirement for a strong inhibitory influence exerted by horizontal cells. The collaterals should allow the horizontal cells to influence the polarization of the bipolar cell membrane at a main dendritic branch, bypassing the trees of thin end branches at the terminals. The large diameter of the collaterals should favor rapid conduction of any potential change generated by the horizontal cells. The extensive connections between small horizontal cells at the collaterals should guarantee that the horizontal cells' influence is based on local conditions of illumination.

The observed circuitry can explain directional selectivity for dark images moving along the north–south axis. When the dark image moves from north to south the small horizontal cells are depolarized and consequently increase their release of hyperpolarizing transmitter at their connections to bipolar cell 1. The connections at the collaterals then establish a hyperpolarization block for the transmission of an off-response of bipolar cell 1. The direction north to south will therefore be the null direction. A movement in the opposite direction will not result in any inhibition of the off-response of bipolar cell 1. The direction south to north is thus the preferred direction. In fact, if we assume that the movement of the dark image from south to north leads to local depolarization of the large horizontal cell processes approaching terminals 1 and 2 from south, the off-response will be facilitated. This follows from the observed multiple connections of the large horizontal cell processes approaching the terminals from the south to small horizontal cell processes approaching from north. The depolarzation of the large horizontal cell process will make this process increase its hyperpolarizing pressure on the small horizontal cells, leading to these cells reducing the hyperpolarizing pressure they exert at the connections to bipolar cell 1 at the synaptic ribbon complexes.

When a bright image moves from north to south, the small horizontal cells are hyperpolarized and consequently reduce their hyperpolarizing pressure on bipolar cell 1, a change that facilitates the on-response.

The connections of bipolar cell 1 in the outer circuitry layer as described here can therefore not account for a directionally selective response to a bright image moving over the retina. We therefore look for other neural connections that can contribute to the circuitry to satisfy this aspect of directionally selective neural responses.

Suppose, for instance, that the intrinsic bipolar cells are involved in such a circuit; connections of bipolar cell 1 to intrinsic bipolar cells furnish structural evidence for this possibility. However, a simple explanation for the generation of a directionally selective response when a bright image moves over the receptive field of bipolar cell 1 is possible without including the intrinsic bipolar cells.

For the bipolar cells transmitting information regarding discrete changes in the input from photoreceptors, we have proposed that both horizontal cells and amacrine cells guard against background noise. If we include the amacrine cells in our circuit, we can easily explain how a bright image moving from north to south inhibits bipolar cell 1, preventing an on-response when the image reaches the center of the receptive field.

When the bright image moves from north to south the small horizontal cells gradually become more and more hyperpolarized. The consequent reduction in the hyperpolarizing pressure exerted by the horizontal cells on bipolar cell 1 at the collaterals and at the synaptic ribbon complexes makes the photoreceptors drive the membrane potential of bipolar cell 1 toward depolarization. This gradual depolarization of bipolar cell 1 makes the amacrine cells increase their hyperpolarizing pressure in a correspondingly gradual fashion with the establishment of a hyperpolarization block of the transmission between bipolar cell 1 and a ganglion cell to which it is connected. The gradual depolarization of bipolar cell 1 and the corresponding gradual increase in hyperpolarizing pressure exerted by the amacrine cells prevents bipolar cell 1 from depolarizing the ganglion cell at a level above the threshold for a ganglion cell response.

Bipolar cell 1 responds with an on-response when the intensity of the light covering the entire receptive field is increased and with an off-response when the intensity is reduced because under these conditions the changes in the intensity of the light are abrupt, not allowing time for the building up of a hyperpolarization block even when the functional receptive fields of the small horizontal cells are involved in the stimulus.

When a bright image moves from south to north no small horizontal cells interfere and bipolar cell 1 transmits a response to the ganglion cell. This direction is therefore still the preferred direction.

4. Electrophysiological Recordings from Bipolar Cells

The electrophysiological recordings of Wyatt and Daw (1975) suggest strongly that the inhibition associated with the movement in the null direction is imposed by horizontal cells. Wyatt and Daw thus showed that if a stationary flashing spot is close to the boundary of the receptive field center on the null side, inhibition can be generated by moving the second spot in the null direction within the adjacent surround. The observed asymmetry of the inhibitory area relative to the stationary flashing spot also agrees with the inhibition being mediated by horizontal cells with an arrangement like that of the small horizontal cells. The limited size of the area from which inhibition could be generated appears to be compatible with inhibition mediated by small horizontal cells.

It is generally believed, however, that the circuit for directional selectivity is located entirely in the inner plexiform layer and that it is the amacrine cells that determine the ganglion cell response on the basis of the input from several bipolar cells (Dowling, 1967; Dowling and Werblin, 1971; Werblin and Dowling, 1969; Werblin, 1970; Norton *et al.*, 1970). These assumptions are understandable because the complex circuitry of the outer plexiform layer has been unknown while the inner plexiform layer, being thicker, impresses one as competent to accommodate complex circuitry.

One argument against bipolar cells being units of this circuitry is that no directionally selective responses have been recorded from bipolar cells. Obviously, this can reflect shortcomings of the electrophysiological technique as well.

As mentioned above, electrophysiological recordings from bipolar cells are technically difficult because the cells are so small. Researchers therefore selected species of animals whose retinal neurons are particularly large, and little attention has been paid to differences in retinal function among species. This is illustrated by the futil efforts of Norton and co-workers (1970) to demonstrate directionally selective responses from bipolar cells in the *Necturus* retina. In later recordings from ganglion cells in the *Necturus* retina by Tuttle (1977), no directionally selective responses could be detected, and according to the recordings of Karwoski and Burkhardt (1976) only a few percent (about 4%) of the ganglion cells in the *Necturus* retina elicited such responses. The *Necturus* retina is therefore unsuitable for an electrophysiological analysis of directionally selective bipolar cells because it requires very extensive sampling of technically difficult recordings from bipolar cells.

Despite this restriction, Norton *et al.* (1970) reported that some units in the inner nuclear layer of the *Necturus* retina respond differently to stimuli moving in opposite directions along several axes across the receptive field. The difference in response for movement in opposite directions was more pronounced along some axes than others. They concluded "This is certainly not specific directional selectivity but it does suggest that a unit at this level of the retinal organization is affected by the distribution and directions of light on the retina." The cells from which the recordings were made were too sensitive to be analyzed extensively, indicating that they were small cells with their function impaired by the electrode.

We can conclude that negative results of recordings from bipolar cells will have no significance until recordings are made from retinas in which directionally selective units have been shown to be abundant and until recordings can be made from all sizes of bipolar cells.

5. Edge Detectors

Edge detectors do not respond to the turning on or off of light when the entire receptive field is illuminated. However, they are activated when an image of a dark object is introduced into the receptive field. Such responses have been recorded at the ganglion cell level and several different types of edge detectors have been described.

It appeared that the connections of bipolar cells to small horizontal cells could establish conditions for edge detection at the level of bipolar cells. These connections transmit information regarding the illuminance over a restricted area of the bipolar cell's receptive field, which appears to be a basis for edge detection. The information transmitted by the small horizontal cells will, in fact, concern not only the presence of an edge but also the location of the edge relative to the receptive field of the bipolar cell.

What characterizes an edge detector is the absence of a response when light covering the entire receptive field is turned on and shut off. It is the introduction of an edge covering a certain area of the receptive field that generates a response. The edge detector must therefore be exposed to strong inhibition, and the introduction of an edge releases the inhibition.

If we now consider the possibility that bipolar cells can function as edge detectors, it is clear that horizontal cells cannot account for the inhibition of the bipolar cell (Chapter 3, Section 22). The outer circuitry layer can therefore not be involved directly in the inhibition of the transmission between the edge detector–bipolar cell and a ganglion cell. This must be a function of the inner plexiform layer, where we proposed that the amacrine cells are capable of controlling transmission between bipolar cells and ganglion cells.

If an amacrine cell is to assist in edge detection, it must block the transmission between the edge detector–bipolar cell and a ganglion cell to exclude ganglion cell responses when light over the bipolar cell's receptive field is turned on or shut off. This requires that the amacrine cell be exposed to a depolarizing pressure sufficiently strong to make it block the transmission between the bipolar cell and the ganglion cell by its release of hyperpolarizing transmitter. The detection of an edge is then possible if the depolarizing pressure exerted on the amacrine cell is reduced at the connection between the edge detector-bipolar cell and the ganglion cell.

The neural interactions within the inner plexiform layer are primarily dependent on the input from the outer circuitry layer mediated by the bipolar cells. We must therefore find a bipolar cell in the outer circuitry layer that can maintain a sufficiently strong depolarizing pressure on an amacrine cell at the synapse between the edge detector–bipolar cell and the ganglion cell and that is exposed to a proper modulation of this pressure.

The type of intrinsic bipolar cells that lacks connections to large horizontal cells at the cone terminals but has connections to small horizontal cells in the subsynaptic neuropil appears suitable for this function. The lack of connections to large horizontal cells means that this bipolar cell maintains a continuous strong depolarizing pressure on the amacrine cells to which it is connected and that only small horizontal cells control this depolarizing pressure. The approach direction of the two small horizontal cells connected to this type of bipolar cell at terminal 1 was from north. The end branches of these horizontal cells made no other connections to neurons in the subsynaptic neuropil, and they were not connected to the terminal. They ended at their contacts at the intrinsic bipolar cell.

We can now propose the following circuit for edge detection. The intrinsic bipolar cell exposes an amacrine cell connected to the edge detector–off-bipolar cell and the associated ganglion cell to a strong depolarizing pressure. As a consequence, the amacrine cell blocks transmission at the synapse between the bipolar cell and the ganglion cell. When a dark image covers the functional receptive field of the small horizontal cells connected to the intrinsic bipolar cell, the depolarization of the small horizontal cells makes them increase their hyperpolarizing pressure on the intrinsic bipolar cell. The consequent reduction of the depolarizing pressure maintained by the intrinsic bipolar cell on the amacrine cell releases the hyperpolarization block maintained by the amacrine cell. When the dark edge moves and eventually covers the center of the receptive field of the edge detector–off-bipolar cell, this neuron generates an off-response that is transmitted to the ganglion cell.

This response requires that the intrinsic bipolar cell becomes hyperpolarized before the dark image moves over the center of the receptive field so that there is sufficient time for the intrinsic bipolar cells to affect the release of hyperpolarizing transmitter by the amacrine cells. This requirement explains the observation that this type of edge detector responds to a dark image moving into the receptive field, whereas it does not respond to a stationary dark image projected onto the same part of the receptive field. In the latter case, the small horizontal cell's influence on the intrinsic bipolar cell is counteracted by the increase in the depolarizing pressure exerted by the photoreceptors on the intrinsic bipolar cell.

Electrophysiological recordings at the ganglion cell level revealed the presence of edge detectors of this kind, but no corresponding studies have been pursued at the bipolar cell level.

Edge detection at this level could contribute importantly to the transmission of integrated information from the retina to the brain and could furnish a basis for the transmission of information regarding image patterns. The very common occurrence of bipolar cells connected to small horizontal cells could allow information transmitted to the brain to be based on certain basic patterns in the distribution of light within the parts of images that are projected onto the bipolar cell's receptive field.

Consider, for instance, the possibility that the small horizontal cells connected to different dendritic branches of a bipolar cell approach the branches from different directions. This organization would allow a bipolar cell to transmit information regarding such patterns. If some dendrites are connected to small horizontal cells approaching from north while other dendrites are connected to small horizontal cells approaching from west, and these photoreceptors are properly distributed in the retina, a bipolar cell could transmit the information that there is a corner with a particular orientation in the image.

At the level of ganglion cells different contributions of edge-detecting bipolar cells could make the ganglion cell transmit information regarding image patterns. Might this be the basis for the reconstruction of the perceived image in the brain? Do the ganglion cells perform the first step in the assembly of patterns from primary data contributed by the bipolar cells?

It is generally assumed that the high acuity of foveal vision in the human eye is due to the established 1 : 1 relationship between bipolar cells and photoreceptors. But each cone is connected to several bipolar cells, so obviously each photoreceptor cannot be connected to the ganglion cells by its own private channel. Each bipolar cell must transmit information from several cones, and it seems reasonable to propose that the information transmitted by each bipolar cell consists of a pattern of light distribution and not information regarding the light intensity at individual points, each of which corresonds to one cone.

Some indication of pattern recognition at the level of the retina has been found in electrophysiological recordings. The "movement dark-convex-edge-detectors" described by Lettvin and co-workers (1959, 1961) may represent an example of such recognition of patterns at the level of the ganglion cells.

Several other types of edge detectors, such as orientation-selective units, local-edge detectors (Lettvin *et al.*, 1959, 1961), and changing-contrast detectors (Maturana *et al.*, 1960) have been identified by recordings from ganglion cells. The latter detectors were identified in the frog retina; the first two were found in the rabbit retina.

Chapter 7: The Duplicity Theory

1. Some Discrepancies in the Duplicity Theory

It is reasonable to expect that the concepts developed when a functional system is analyzed will vary with the type of information that the methodology has contributed. The duplicity theory was developed from a comparison between the ratios of rods to cones in retinas of animals active in full daylight and animals active in dim light. The retinas of diurnally active animals contain more cones than the retinas of nocturnally active animals (Schultze, 1866). This led to the conclusion that cones are designed for vision in daylight, photopic vision, while rods are designed for vision at night, scotopic vision. This concept has become a dogma that has been accepted as a satisfactory explanation for the presence of two types of photoreceptors in the mammalian retina.

Later studies on human vision revealed a physiological duplicity of vision that was explained by differences in the function of rods and cones. The discovery of a change in the spectral sensitivity of the retina when the illuminance is decreased, the Purkinje shift, was explained by the two types of photoreceptors with different spectral sensitivity functioning at high and low illuminance, respectively. The observation that during dark adaptation the threshold for cones decreases faster than that for the rods showed another functional difference between rod vision and cone vision, and this difference is further manifested in the difference in the critical flicker frequency, which is considerably lower for rod vision than for cone vision. Rods have been considered to be slower-reacting photoreceptors than cones.

The association of rods with scotopic vision was explained by their higher sensitivity. They were estimated to be a thousand times more sensitive than cones. This was considered to be a consequence of a difference in the concentration of photopigment in the two types of photoreceptors. It was concluded that there was an extremely low concentration of photopigment in cones (Wald, 1954).

A detailed analysis of rod and cone vision has shown, however, that many of the phenomena attributed to different functions of rods and cones must have other explanations. The difference in threshold of rods and cones was shown to decrease with a reduction in the size of the area on the retina covered by the test stimulus (Baumgardt, 1949). When, in addition, corrections for the difference in the density in the distribution of rods and cones over the surface of the retina were introduced, it turned out that the thresholds for rods and cones were essentially the same (Weale, 1958).

A similar sensitivity of rods and cones requires that both types of photoreceptors have a similar ability to capture photons. This in turn means that the effective concentration of photopigment should be the same in the two types of photoreceptors. Measurements have shown that this is true (Weale, 1961). Therefore the maintainance of rod vision and the shutting off of cone vision at a low illuminance is unlikely to be due to a difference in the sensitivity of the photoreceptors.

The duplicity of the retina has been well established physiologically. Structurally, there are two types of photoreceptors in the mammalian retina. It seems reasonable to link the functional duplicity to the structural duplicity. However, the question is *how* to link function to structure. If there is little difference in the thresholds of rods and cones, there is no basis for a duplicity theory that requires an about thousandfold difference in the threshold.

In a lecture in 1959, Weale (1961) exposed the duplicity theory of vision to a critical examination and concluded that most of the characteristic differences of rod and cone vision were due not to basic differences in the properties of the photoreceptors but to neural factors. He expressed this point of view in the following way: "The rods and cones are the aerials: the work is done by the power supply and the amplifiers behind them."

2. The Temporal Resolution of Bipolar Cell Responses

The bipolar cell responses deduced in Chapter 3 offer an explanation why these cells respond differently at low and high illuminance. These deductions were based on the horizontal cells exerting a hyperpolarizing pressure opposing a depolarizing pressure maintained by the photoreceptors. At a low illuminance, when the horizontal cells approach maximal depolarization, the hyperpolarizing pressure is maintained at a more or less constant level, while at a higher illuminance this pressure is modulated. This modulation changes otherwise sustained depolarization or hyperpolarization of the bipolar cells to transient potential changes. The modulation of the responses must speed up the return of the membrane potential to a level of polarization from which a new full-amplitude response can be evoked, improving the temporal resolution of bipolar cell responses. The critical flicker frequency should therefore be higher when the horizontal cells can modulate the bipolar cell responses than when they maintain a more or less constant hyperpolarizing pressure opposing the depolarization exerted on the bipolar cells by the photoreceptors.

At low illuminance the hyperpolarizing pressure exerted by the horizontal cells on the bipolar cells remains almost constant within a rather wide range of illuminance (Chapter 3, Section 9). Within this range we expect the critical flicker frequency to be constant. When the illuminance reaches a level at which the horizontal cells start to modulate the hyperpolarizing pressure sufficiently to affect the bipolar cell responses, the critical flicker frequency should increase. As the illuminance increases further and the amplitude of the modulation of the hyperpolarizing pressure becomes larger, the critical flicker frequency should increase with increasing illuminance until the modulation has reached the maximum amplitude. The critical flicker frequency should then level off. Eventually it should decrease when the horizontal cells approach maximal hyperpolarization, causing a decrease in the amplitude of the modulation of the hyperpolarizing pressure.

It can also be deduced that the connections between cones and bipolar cells connected to rods lower the critical flicker frequency associated with rod vision below that of cone vision. This follows from the cones opposing the potential changes imposed on the bipolar cells connected to rods when both rods and cones are stimulated by light. With two opposing influences acting simultaneously on the bipolar cell, not only the amplitude but also the speed of potential change will be affected.

Another factor that must affect the critical flicker frequency is the precision with which the flicker phenomenon is detected. This becomes a factor that interferes with the determination of the critical flicker frequency when the intensity of the light stimulus approaches the thresholds for rod vision and for cone vision.

This deduced relationship between critical flicker frequency and illuminance agrees with the observed relationship from the study of Hecht and Smith (1936) shown in Fig. 7.1. The deduction also agrees with the observation that for small test areas the critical flicker frequency increases when the surround is illuminated (Berger, 1953, 1954). Illumination of the surround pushes the membrane potential of the horizontal cells toward the range within which the horizontal cells' modulation of the hyperpolarizing pressure becomes effective.

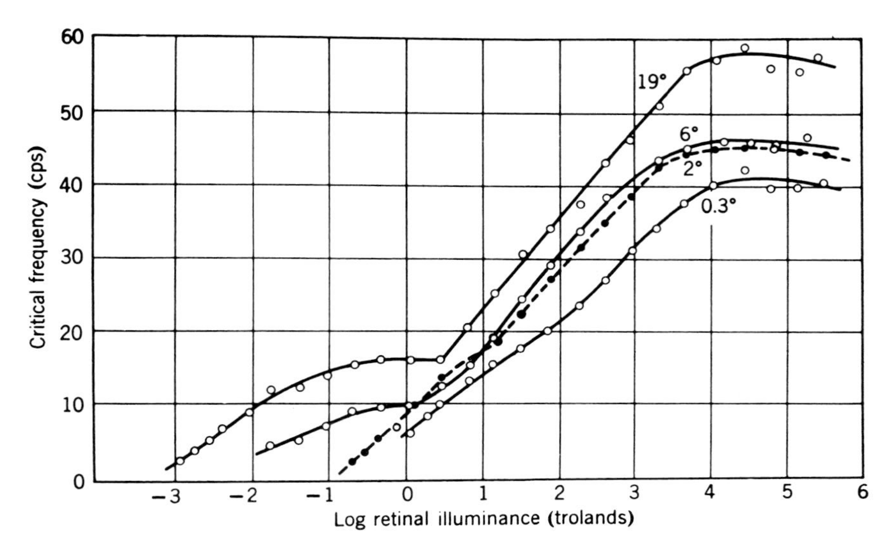

Fig. 7.1. Critical flicker frequencies as a function of log retinal illuminance determined for centrally fixed test fields of different sizes. With a test field of 19° both rods and cones are stimulated, and the two components contributed by rods and cones are clearly shown. At the smallest test fields, 0.3° and 2°, the stimulation involves mainly cones, and no rod component can be discerned. (Hecht and Smith, 1936. Reproduced from the *Journal of General Physiology*, 1936, vol. 19, pp. 979–989 by copyright permission of the Rockefeller University Press.)

Hecht and Shlaer (1936) determined the critical flicker frequency for lights of different wavelengths (Fig. 7.2), showing that rod vision accounts for the plateau at low illuminance while cone vision accounts for the frequencies at higher illuminance. This association of the different parts of the curve with rod vision and cone vision has been one basis for the conclusion that rods and cones function differently. However, no simple explanation for the shape of the curve relating critical flicker frequency to illuminance has been found.

The connections between the cones and the bipolar cells connected to rods explain that at high illuminance the critical flicker frequency is associated with cone vision. That rod vision accounts for the critical flicker frequencies at low illuminance is explained by the known fact that the relative thresholds of rod and cone vision are reversed at low illuminance. However, it remains to be explained how this reversal is accomplished. The critical flicker frequencies as a function of illuminance offer information that, combined with our deductions, can explain the reversal.

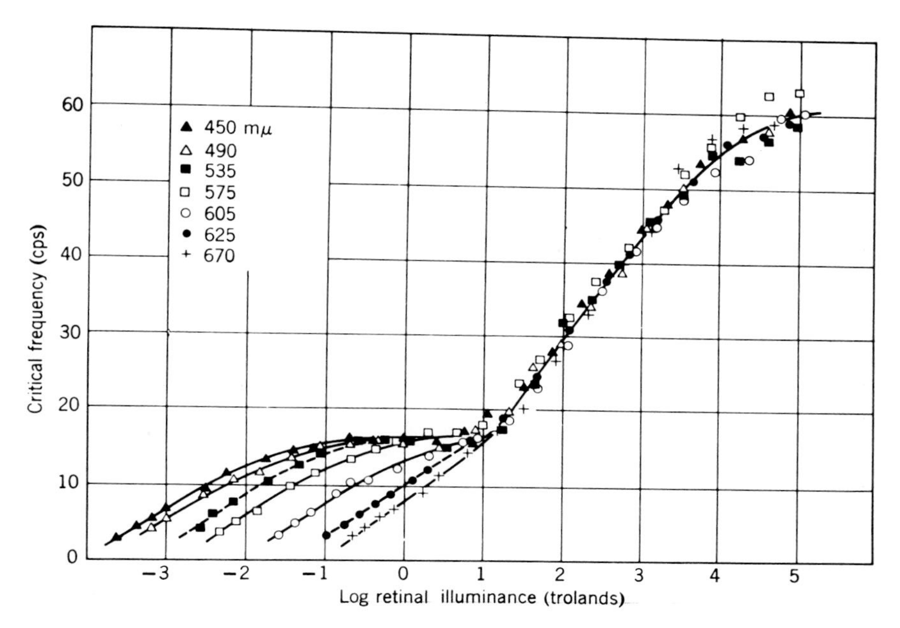

Fig. 7.2. Critical flicker frequency as a function of log retinal illuminance for seven spectral regions. (Hecht and Shlaer, 1936. Reproduced from the *Journal of General Physiology*, 1936, vol. 19, pp. 956–979 by copyright permission of the Rockefeller University Press.)

Thus, it is obvious from Fig. 7.2 that the reversal to lower threshold for rod vision relative to cone vision takes place at an illuminance at which the curve levels off. According to our deduction this corresponds to the transition from the range of illuminance within which horizontal cells modulate bipolar cell responses to the range where such modulation is absent. We can therefore relate the rise of the threshold for cone vision relative to that of rod vision to the reduction and elimination of the modulation of the bipolar cell responses by the horizontal cells, that is, to the illuminance at which the horizontal cells maintain a strong hyperpolarizing pressure on the bipolar cells. This relationship can explain the change in relative thresholds of rod and cone vision at a low illuminance, as will be discussed in Section 5 of this chapter.

The curve showing the observed critical flicker frequencies at 670 nm, which are accounted for by cone vision, reveals a rapid decrease in the critical flicker frequencies at an illuminance below that at which the curves for other wavelengths level off. There is a knick in the 670-nm curve, indicating that at this illuminance the critical flicker frequency is related to a factor different from that determining the shape of the curve

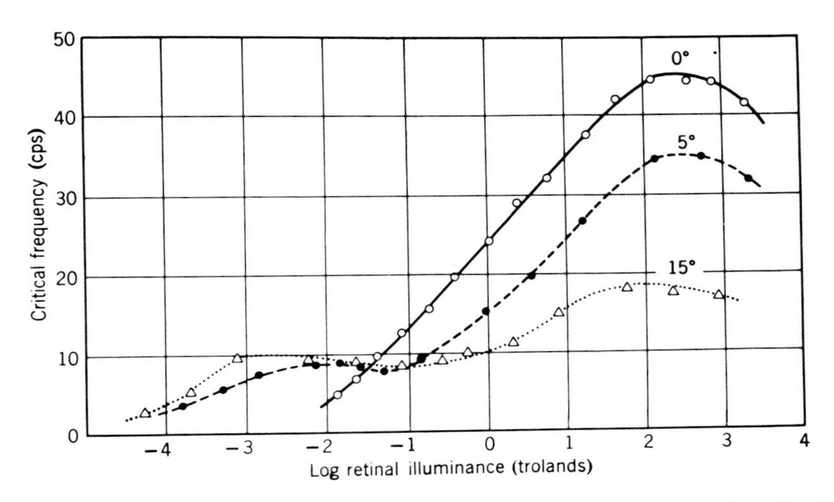

Fig. 7.3. Critical flicker frequency as a function of log retinal illuminance for white light for three different locations: at the fovea and 5° and 15° above the fovea. (Hecht and Verrijp, 1933. Reproduced from the *Journal of General Physiology,* 1933, vol. 17, pp. 251–265 by copyright permission of the Rockefeller University Press.)

above the knick. It appears likely that the slope below this point is determined by the uncertainty by which the flicker phenomenon can be detected. This would agree with a rapid rise in the threshold for cone vision at illuminances below this point.

We explained that cone vision accounts for the critical flicker frequency at a high illuminance because the cones raise the threshold for rod vision at these intensities. It is obvious that the observed critical flicker frequencies must be accounted for by the system with the lowest threshold and that the observation does not allow any conclusions whether rod vision is active or not.

According to Aguilar and Stiles (1954), rods contribute to vision at high illuminance and are shut off only at extreme intensities of light, as pointed out in Chapter 5, Section 3, and according to our deduction the critical flicker frequency for rod vision should be lowered as a consequence of the inhibitory interactions of the cones on bipolar cells connected to rods. It now appears possible that the critical flicker frequencies determined by the use of test light covering an area of the retina in which the rods dominate over cones may reveal the critical flicker frequencies of rod vision at illuminance where the bipolar cells connected to rods are inhibited by cones.

Hecht and Verrijp (1933) determined the critical flicker frequencies with a test light covering an area 15° above the fovea (Fig. 7.3). The density of rods here is almost at its maximum, and rods are about three times more abundant than cones. It is therefore possible that rod vision accounts for the critical flicker frequency within this area. This frequency was less than half the frequency determined with a test light covering the fovea. This considerable difference between the critical flicker frequencies determined for the fovea and those determined for an area 15° above the fovea is easier to explain by rod vision accounting for the latter frequencies than by cone vision differing for foveal cones and extrafoveal cones.

The amplitude of the bipolar cell responses should vary with the frequency of the flicker. As the critical flicker frequency is approached, the amplitude should correspond to the threshold amplitude for the transmission of signals by the bipolar cells activating the ganglion cells. As the frequency is lowered, the amplitude should increase, corresponding to the increase in time available for pushing the membrane potential of the bipolar cells toward a baseline potential. The amplitude of the response of the bipolar cells to the first stimulus when the flickering light is turned on should be larger than the amplitude of the following responses. This agrees with the electroretinographic recordings.

3. Adaptation

The photoreceptors in the retina can increase their response amplitude with increasing intensity of light stimulation over a certain limited range. The smaller the change in the intensity of the stimulus required to produce a change of a certain magnitude in the response amplitude, the higher is the contrast sensitivity or brightness discrimination. A curve showing the relationship between the logarithm of light intensity and the amplitude of the response has a middle steep part that represents the range of highest contrast sensitivity and, consequently, the range of highest brightness discrimination. It extends over a range of about one log unit of light intensity. This range corresponds to the range of light intensities that are reflected by various objects at practically any level of ambient illuminance.

The range of the highest contrast sensitivity is, however, narrow in comparison to the range of variation in the levels of ambient illuminance at which intensity discrimination is possible. The latter range extends over more than seven log units, equivalent to a change in illuminance from moonlight to bright sunlight. This extension of the range is due to adaptation of the photoreceptors to various levels of ambient illuminance by variation of their photon-catching capacity. The adaptation involves a translocation of the range of optimal brightness discrimination with changes in ambient illumination. In this manner the brightness discrimination is maintained practically constant over the entire range of adaptation.

The range of optimal brightness discrimination corresponds to a range within which the smallest increments in the intensity of the illuminance cause increments in the amplitude of the bipolar cell membrane potential that after transmission to the brain by the ganglion cells exceed the threshold for discrimination by the visual centers in the brain.

We can therefore relate the brightness discrimination to the amplitude of the depolarization of the bipolar cells, which depends upon the extent to which the horizontal cells modulate the changes in the membrane potential. We expect that the range of highest brightness discrimination coincides with the range within which the horizontal cells modulate the bipolar cell responses most extensively. This modulation becomes less extensive as the illuminance is lowered until at low illuminance it vanishes. It also becomes less extensive at a high illuminance when the horizontal cells approach maximal hyperpolarization. On this basis, therefore, we can deduce that the range of highest brightness discrimination corresponds to an intermediate range of illuminance, in agreement with the observations.

The photon-catching capability of the photoreceptors is regulated by the bleaching and regeneration of the photopigments. When we move from bright sunshine to deep shade the regeneration of bleached photopigments makes the threshold for stimulation of the photoreceptors decrease. However, this decrease is preceded by a rise in the threshold during a very short period (Crawford, 1947). After this rise the threshold decreases rapidly at first and then at a slower rate and then levels off. It then decreases rapidly again to reach a minimum after 30 min to 2 h (Fig. 7.4).

Dark adaptation thus proceeds in two phases. The first phase is accounted for by cone adaptation and the second phase is accounted for by rod adaptation. The course of dark adaptation as revealed by this two-phase change in the threshold has been interpreted as showing that the adaptation of the rods is delayed relative to that of the cones and that this delay is longer the higher the intensity of the preadapting light (Hecht *et al.* 1937). The rate at which the rod threshold decreases also slows with increasing intensity of the preadapting light. This difference has been thought to reveal an additional functional difference between rods and cones with cones adjusting faster to changes in the illuminance than do rods.

As a consequence of the adaptation being determined on the basis of thresholds, the adaptation of rods cannot be studied as long as the intensity of the test light is high enough for the cones to inhibit bipolar cells connected to rods because these connections then raise the rod

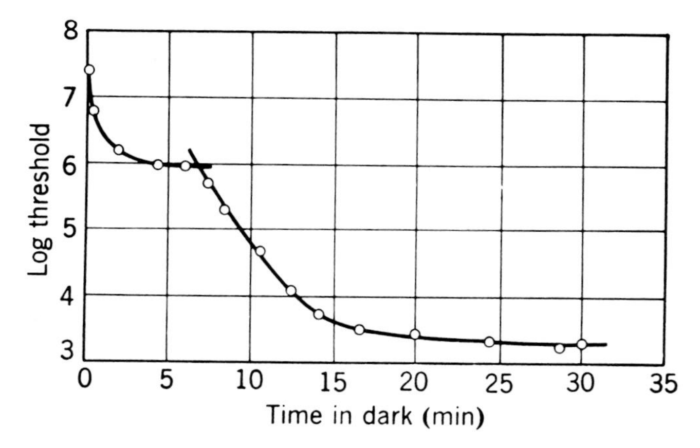

Fig. 7.4. Change in the threshold for vision during dark adaptation when measured within an area located outside the fovea. (Data from Hecht and Shlaer, 1938. From Graham. Reprinted by permission of John Wiley and Sons from "Vision and Visual Perception," 1966, Fig. 4, 10, p. 75.)

threshold above that of cone vision. Not until the intensity of the test light is below this level can the threshold of rod vision be determined.

Obviously, rhodopsin must regenerate from the onset of the dark adaptation, and the adaptation of rod vision may follow a course corresponding to that shown by extrapolating the rod adaptation curve to the ordinate. That the curve then crosses the ordinate above the threshold for cone vision is explained by the cones raising the threshold for rod vision.

The observations do not allow any conclusions regarding a delay in the adaptation of rods to be drawn, and they cannot be used as an indication that there is a basic difference in the way rods and cones adapt. However, the possibility that cone photopigments regenerate at a rate differing from that of rhodopsin cannot be ruled out.

The transient early rise in threshold during dark adaptation (Fig. 7.5) can be explained as a consequence of the neural connections of the photoreceptors. The transmission of signals by the on-bipolar cells requires that the hyperpolarizing pressure exerted by the horizontal cells be reduced sufficiently to allow the photoreceptors to depolarize the on-bipolar cells. When the light is turned off after a period of preadaptation, the horizontal cells become maximally depolarized and block the transmission between photoreceptors and bipolar cells. During this phase we would therefore expect the threshold to increase.

As the hyperpolarizing pressure exerted by the large horizontal cells

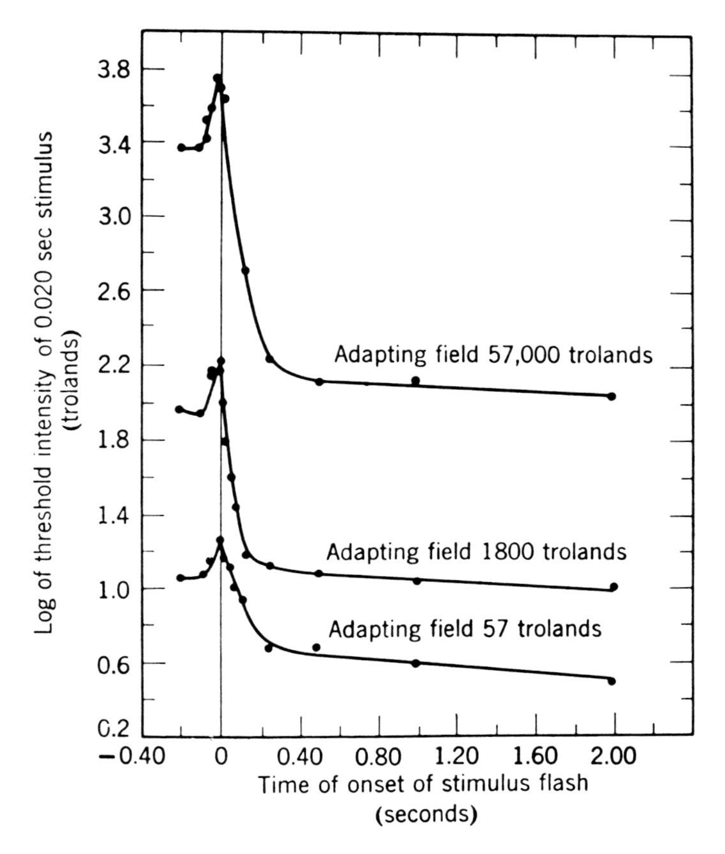

Fig. 7.5. The earliest phase of dark adaptation with an increase in the threshold for stimuli applied just before and after the cessation of the preadapting light. The stimulus consisted of a brief flash of light. The preadapting light was shut off at time zero. From Baker (1953).

affects the small horizontal cells, the latter cells reduce their hyperpolarizing pressure at the photoreceptor–bipolar connections and the photoreceptor can now break the hyperpolarization block at the synapse. The threshold is then lowered.

This means that there are two components of dark adaptation, one neural and one associated with the regeneration of the photopigments. The increase of the threshold is confined to a very short period (Crawford, 1947; Baker, 1953, 1963) after the shutting off of the preadaptation light. The threshold appears to increase even before the light has been shut off (Fig. 7.5).

The time during which the threshold is increased corresponds to the time required for the transmission from the photoreceptors to the horizontal cells, the transmission from the large horizontal cells to the small horizontal cells, and the transmission from the small horizontal cells to the photoreceptors and the bipolar cells. This means that there are three synaptic delays during which the threshold of the bipolar cells is raised. The rise starts when the photoreceptors depolarize the horizontal cells, that is, during one synaptic transmission. The stimulation of an on-bipolar cell requires that the horizontal cell's hyperpolarizing pressure on the photoreceptor and on the bipolar cell be reduced. The photoreceptor can then depolarize the bipolar cell. Therefore this occurs after more than two synaptic delays.

When a test stimulus is applied at a time corresponding to two synaptic transmissions before the preadaptation light is shut off, the threshold is raised because at the time the evoked response of the photoreceptors should lead to a depolarization of the bipolar cells the effect of the shutting off of the preadaptation light will increase the hyperpolarizing pressure exerted by the horizontal cells.

This neural component should not be confused with the "neural component" of dark adaptation that has been assumed to correspond to the fast phase in threshold reduction (Baker, 1953, 1963; Baker *et al.*, 1959; Wald, 1961; Dowling, 1963). The neural nature of this phase in dark adaptation has been questioned by Rushton and Powell (1972), who showed that the Bunsen–Roscoe law applied to dark adaptation. The adaptation during this fast phase must therefore be due to a photochemical mechanism and not to a neural influence.

4. Rod and Cone Thresholds

Baumgardt (1949) and Arden and Weale (1954) concluded that rod and cone thresholds are similar. The observations on which this conclusion is based involved measurements of rod and cone thresholds with small test fields. According to Baumgardt the difference in the thresholds decreased when the size of the test field was reduced, and an extrapolation to very small fields indicated that there was no great difference in the thresholds.

Arden and Weale, comparing the threshold for the fovea and for the periphery of the retina with small test fields, found that the absolute thresholds for rods and for cones were almost the same.

That the observations of Baumgardt and of Arden and Weale showed rods and cones to have the same threshold was questioned by Haig (1958) and Pirenne (1962). However, the observations clearly revealed a dependence of the thresholds on the size of the test field.

The neural connections of rods and of cones exclude determinations of the photoreceptor thresholds without recording from the photoreceptors. The lateral inhibition responsible for spatial brightness contrast enhancement makes the threshold change with the size of the test field and efferent bipolar cells which are connected to cone terminals but not to rod terminals are likely to affect the transmission of signals at cone terminals, and it is likely that the efferent bipolar cells exert an inhibitory influence.

Threshold determinations will involve rod vision and cone vision and not the photoreceptors. Thus the most reliable information is contributed by the measurements of the photopigment concentration in rods and cones. These measurements have not revealed any difference that can account for the extensive difference in photon-catching capability required by the duplicity theory.

5. Is the Cone Threshold Regulated by the Illuminance?

Let us consider the possibility that the high threshold for cones at a low illuminance is the consequence of neural inhibition of the cones. As pointed out above, the curve showing the critical flicker frequencies as a function of illuminance leveled off at an illuminance at which the cone threshold increased relative to the threshold for rod vision. According to our deduction, this leveling off occurs when the illuminance reaches the level at which the horizontal cells can no longer modulate the bipolar cell responses. That is, the cone threshold rises at the same illuminance at which the horizontal cells exert a maximal hyperpolarizing pressure on photoreceptors and on bipolar cells.

This we interpret as suggesting a relationship between horizontal cell influence and cone threshold. Such an influence cannot be exerted by horizontal cells directly because they are connected to rods and cones in the same way. The efferent bipolar cells, on the other hand, qualify for this function. There is always one efferent bipolar cell connected to cones but efferent bipolar cells are never connected to rods. The efferent bipolar cells are characterized by their deeply invaginated endings filled with synaptic vesicles. They can expose the cones to a depolarizing pressure that will affect the transmission of signals at the cone terminals, and the endings are located in such a way that they can interfere with the signal transmission at the terminal.

This is illustrated by the connections of the efferent bipolar cell (14) at terminal 1. It contacted horizontal cell endings through lateral connections at 4 of the 16 synaptic ribbon complexes. All connections involved small horizontal cells. This seems to be significant because at a low illuminance the small horizontal cells regulate the gate controlling the transmission of signals from the photoreceptors to the on-bipolar cells. In addition, bipolar cell 14 contacted several bipolar cells directly, for instance, the core bipolar cell (39) and bipolar cell 13 that contributed endings to five synaptic ribbon complexes.

The connections of the efferent bipolar cell at terminal 1 is selective and do not involve all neurons connected to the terminal.

Any inhibitory action of the efferent bipolar cells must be regulated mainly by their connections in the inner plexiform layer. The amacrine cells can here expose them to a hyperpolarizing pressure strong enough to block any inhibitory action. For activation of the efferent bipolar cells, intrinsic bipolar cells not connected to horizontal cells may be involved, and the blocking of the efferent bipolar cells then depends on a balance between the hyperpolarizing pressure exerted by the amacrine cells and the depolarizing pressure exerted by these intrinsic bipolar cells on the efferent bipolar cell.

The hyperpolarizing pressure exerted by the amacrine cells is regulated by the overall input from afferent bipolar cells that maintain a constant depolarizing pressure on the amacrine cells. When this pressure is high enough, the amacrine cells may maintain a hyperpolarization block on the efferent bipolar cells.

When the horizontal cells approach maximal depolarization at a low illuminance and consequently expose the bipolar cells and the photoreceptors to a maximal hyperpolarizing pressure, we can expect that the bipolar cells reduce the depolarizing pressure on the amacrine cells. In contrast, the intrinsic bipolar cells that are not connected to horizontal cells will increase the depolarizing pressure they exert on the efferent bipolar cell as the photoreceptor becomes more depolarized. The balance between hyperpolarizing and depolarizing pressures exerted on the efferent bipolar cell then changes in favor of depolarization with a consequent rise of the threshold for signal transmission at the connections that are under efferent bipolar cell influence at the cone terminal.

In this manner it is possible, hypothetically, to link the leveling off of the critical flicker frequency and the rise in the cone threshold as caused by the horizontal cells approaching maximal depolarization, and it is the change in the horizontal cell influence from a modulated hyperpolarizing pressure to a constant and high hyperpolarizing pressure that creates a distinct difference in retinal function at low and at high illuminance. This difference in horizontal cell influence may furnish the basis for the difference between scotopic and photopic vision.

The primary function of the efferent bipolar cells is unlikely to be to cut off cone vision at a low illuminance, the usefulness of which seems highly questionable. Instead, the efferent bipolar cells are likely to fulfill a function associated with the processing of information at the cone terminal. The selectivity of the connections of the efferent bipolar cell at terminal 1 agrees with this interpretation. The inhibition at a low illuminance may then be a secondary effect and a consequence of the connections the function of which is to allow the efferent bipolar cell to contribute to information processing.

6. The Purkinje Shift

The change in spectral sensitivity of vision when the illuminance is lowered has been thought to reveal that rods function at a low illuminance and cones function at a high illuminance. This Purkinje shift shows that at a low illuminance the spectral sensitivity of vision corresponds to the absorption spectrum of rhodopsin after due corrections. At higher illuminance the spectral sensitivity is pushed toward somewhat longer wavelengths. There is a considerable overlap of the spectral sensitivities at high and at low illuminance. The longer wavelengths characterize the spectral sensitivity of the fovea, and the change has therefore been explained by a change from rod vision to cone vision.

The Purkinje shift is based on the determination of thresholds and allows only the conclusion that there is a change in the relative thresholds of rod and cone vision with illuminance. The observation does not indicate that rod vision is shut off at high illuminance or that cone vision is shut off at low illuminance. Although the observation is compatible with the duplicity theory, it does not prove the theory.

7. A Different View of the Duplicity of Vision

One way to free oneself from the intellectual handicap of knowing too well the generally accepted concepts regarding vision is to assume complete ignorance but to accept as worthwhile a project to design a retina from scratch. The design should satisfy an animal's practical interests: finding and identifying food and recognizing enemies. Thus the fundamental aim of our design is to create vision for the discovery and recognition of objects.

Objects are defined first by their outlines. An enhancement of the outlines to the extent that the brightness distribution at boundaries between objects and their background is distorted would be in the interest of the animal.

The actual design of the retina satisfies this principle. The predominance of neurons that transmit information regarding changes in the illuminance by on-, off-, and on/off-responses is a feature of the design of the retina that favors the detection of boundaries and of changes in image patterns. The spatial brightness contrast enhancement is obviously an important mechanism that facilitates the detection of boundaries in image patterns by exaggerating the brightness contrast at boundaries. The introduction of color discrimination adds further to the identification of objects by enhancing boundaries between fields of different colors.

The design of the retina thus satisfies the interest of the animal with respect to information gathering. These interests are universal, independent of the environment in which the animal lives and of the time of the day at which it hunts for food. The interest of the animal would be best served if the visual system were designed so that the animal can see well over the widest possible range of illuminance.

In our design spatial brightness contrast enhancement can be achieved by direct lateral inhibition by photoreceptors of the transmission of signals from other photoreceptors. It requires that the photoreceptors be connected to the neurons that transmit the signals from the latter photoreceptors. These connections complicate the structure of the photoreceptors by requiring that they be equipped with processes extending laterally. This design requires a considerable expenditure of neural matter which is reduced by the restriction of these processes to only part of the population of photoreceptors. Thus the retina that we design will be equipped with two types of photoreceptors.

The inhibition exerted by photoreceptors risks reducing the range of illuminance over which vision functions. Because only part of the population of photoreceptors can be inhibited, these risks are eliminated.

The enhancement of spatial brightness contrast is improved if the number of inhibitory photoreceptors in the population of photoreceptors is increased. The improvement applies first to the amplitude of the distortion of the real brightness distribution but it also affects the width of the boundary zone within which the contrast is exaggerated.

We therefore expect that within the part, where the retina is designed for optimal conditions of vision, all or practically all photoreceptors are of the type that inhibits the transmission of signals from surrounding photoreceptors. This type of photoreceptor should dominate this part, with the exclusion of the other type.

The changing of the absorption spectrum of the photopigment does not require any change in the basic design of the photoreceptors or of the retina. It requires only channels for the transmission of signals coding for color and a circuit that allows exploitation of these signals.

We then find that the presence of rods and cones in the real retina satisfies the basic requirements for the design of a useful visual system. We would have had difficulty justifying a design with two types of photoreceptors of equal sensitivity, one type functioning in bright light and the other type in dim light. Such a design is nonsensical, particularly in the case of the human eye, in which about 95% of the photoreceptors would not function most of the time!

Let us turn our attention to a step in the visual process after the photoreceptors have responded to the images projected onto the retina: the processing of the information picked up by the photoreceptors. We refer to this information as primary information.

Our first requirement when designing neural tissue for information processing is that the processing occurs with a minimum of loss of the resolution characterizing the primary information. Our second requirement is that, in agreement with a general principle in the designs by Nature, the construction be carried out in the most economical way with respect to the use of neural matter.

The greediest design is one in which the first stage in information processing occurs at the photoreceptors. Any displacement of this function away from the photoreceptors entails a waste of neural matter and risk of a loss in the resolution of the primary information.

The processing of information requires that primary information, collected over a certain area of the retina, be integrated. This area we refer to as the receptive field of information processing. Integration of primary information requires that this information be collected at a center, referred to as the primary center for information processing. The smaller the receptive field required to feed sufficient information to this center, the closer the centers can be arranged without the receptive fields overlapping and the higher will be the resolution at which the processed information can be obtained. If the receptive fields overlap, the resolution can be improved further. If the receptive fields do not overlap, the primary information processing centers will be spaced more or less widely apart.

The primary information supplied by the photoreceptors can be processed in various ways, allowing different types of integrated information to be extracted. Because these various modes of information processing utilize primary information supplied from the same area at the retina, we can save neural matter if we let one center process all combinations of pieces of information.

Locating this center at the terminal of a photoreceptor allows the shortest possible path for the transmission of primary information to the center. This means a saving in neural matter because it affects not only the length of the connections but also their required diameter. We therefore achieve the most economical design by means of short and thin processes. The shortness also guarantees that the integration of information takes a minimum time. A further gain is the synchronization of the signals through the channels away from the information processing center by the triggering action of the photoreceptor at which the center is located.

One way to make the center compact is to eliminate the insulation of neural connections by interposed processes of insulating cells. Insulation is required to prevent interference with transmission at neural connections. By locating part of these connections in invaginations of the neural membrane of the photoreceptor terminal we can avoid the need for insulating cells in the center.

Such a design requires that a large surface area be available for the connections at the terminal. The photoreceptors that we had equipped with lateral processes offer a large surface area that can be increased further by deep invaginations. We therefore locate the primary information processing centers at these terminals.

Our design of the retina is characterized by two types of photoreceptors, rods and cones, with cone referring to the shape of the terminal. The cones are separated by a distance that can vary depending on the resolution required for primary information processing and for spatial brightness contrast enhancement. The rods fill in between the cones, and they may or may not dominate quantitatively. We can distinguish between the information processing functions of the two types of photoreceptors in the following way. The cones are centers for more or less complex information processing; the rods supply information to the information processing centers through horizontal cells and bipolar cells. They are therefore information-supplying components, and the retina can be considered to be a two-dimensional keyboard on which the projected images activate different combinations of photoreceptors like a pianist's chords. It seems highly likely that this basic principle of neural organization applies to other neural centers with the input to the centers being received in a corresponding "keyboard" layer of neurons.

The difference between the contributions to information processing of rods and cones is illustrated in Fig. 7.6 in a highly schematic way by showing the number of neurons connected to rod and cone terminals. This figure illustrates an aspect of rod and cone function entirely different from the functions usually associated with the two types of photoreceptors.

We have here presented a basis for the functional duplicity of the retina that differs from that stated by the duplicity theory. Instead of distinguishing between a photoreceptor that functions in bright light from

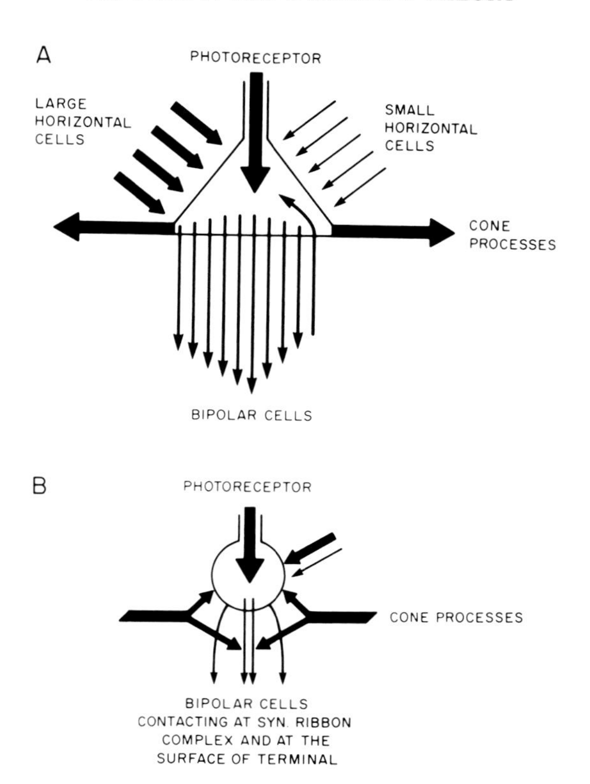

Fig. 7.6. A schematic comparison of neural connections at cone (A) and at rod (B) terminals.

another type that functions in dim light we distinguish between two types of photoreceptors that structurally satisfy the requirements for the enhancement of spatial brightness contrast and for information processing.

The various observations that have been the basis for the concept that rods and cones are functionally very different photoreceptors can be explained by their neural connections. That rod vision is cut off at very high illuminance can be explained by the cones increasing the threshold for rod vision, eventually blocking rod vision without blocking the input from rods to horizontal cells. The rise of the cone threshold and eventual blocking of cone vision at a low illuminance is explained by neural inhibition of cone vision.

The changes of the thresholds during dark adaptation are influenced greatly by the neural connections that make the observed cone and rod thresholds appear to change in very different ways. However, the observations do not reflect the real course of rod adaptation and therefore do not allow the conclusion that there is a basic difference in rod and cone adaptation.

The difference in the critical flicker frequencies of rod and cone vision can also be explained by the neural connections, and there are no reasons to assume that they reflect any differences in the function of the two types of photoreceptors.

The Purkinje shift that has been interpreted as revealing a distinct difference in the range of illuminance at which rods and cones function has been deduced to be a consequence of the cones increasing the threshold for rod vision. The fact that the Purkinje shift is based on the determination of thresholds means that this shift does not allow any conclusions as to whether rods and cones function or do not function at certain ranges of illuminance.

The fundamental difference between rods and cones is the cone processes that make the cones interfere with the transmission of signals from the rods. The neural connections in the retina, not any basic differences in the function of the photoreceptors constitute the real basis for the duplicity of vision. This is an entirely new concept.

As pointed out above, the concepts developed in the analyses of biological systems will be determined by the type of information that the research method has supplied. The information regarding the basic design of the circuitry of the retina obviously differs greatly from the information on which Schultze (1866) based his concept. However, given the criteria used by Schultze to identify cones, there is no reason to assume that his concept that cone-rich retinas characterize diurnal animals and that rod-rich retinas characterize nocturnal animals is wrong. Why then the discrepancy between Schultze's concept and the concept presented here?

One of the major characteristics of photoreceptor structure that Schultze used to identify cones was the conical shape of the outer segment. Later, the different shapes of the terminals, "pedicles," and "spherules" were introduced as further structural features distinguishing rods from cones (Ramon y Cajal, 1894, 1933; Greff, 1900; Pütter, 1912; Rochon-Duvigneaud, 1934; Detwiler, 1943). It is, however, important to emphasize that this structural criterion was not used by Schultze when identifying cones.

Typical conically shaped outer segments are considerably shorter than cylindrical outer segments, and they are on the whole smaller. The differences in the shape and size of the outer segment reflect quantitative differences with respect to the capacity to catch photons. The small outer segments of this type of cone therefore correspond to a threshold for stimulation higher than that characterizing cones with large outer segments, such as the cones in the human fovea. As a consequence, cone vision in animals with cones of this type functions only at a high illuminance, and the brightness contrast-enhancing mechanism functions only in daylight. As the light is reduced, the quality of vision deteriorates, as a consequence of both the elimination of spatial brightness contrast enhancement and the small population of rods. An animal with such a retina must hunt for food in daylight. This does not appear to be an adaptation for daylight living habits but an evolutionary variation that makes the animal unfit for hunting at night.

Schultze's observations apply to animals in which the retinas contain one particular type of cone. These cones are entirely different from human cones, and experiments pursued on human cone vision are therefore not applicable to the cones identified as such by Schultze with respect to threshold for stimulation. In fact, no correlation between photoreceptor types and diurnal and nocturnal living habits has been established for cones when they are defined on the basis of the shape of the terminal. Instead, as pointed out in Chapter 8, Section 2, there are many observations that do not fit with any such correlation.

An analysis of the circuitry of cone-rich retinas defined according to Schultze's criteria could contribute information regarding functional advantages of such retinas. They may offer excellent conditions for color discrimination, for spatial brightness contrast enhancement at a high resolution, and for primary information processing on the basis of small receptive fields for the primary information processing centers. The latter would allow information processing at a high resolution. We might find correlations of much greater interest than the correlation found by Schultze showing gains that may compensate for the impaired vision at a low illuminance.

Chapter 8: Some Conclusions

1. Bipolar Cells as Information Channels

The bipolar cells function as channels through which processed information is transmitted between the outer circuitry layer and the inner plexiform layer. The processing preceding the transmission of information to the latter layer represents a primary step in the integration of information supplied by the photoreceptors. For each afferent channel a particular combination of pieces of information is selected for processing by the pattern of neural connections furnishing the input to the channel. We can therefore distinguish between different modes of information processing associated with the channels.

The patterns of neural connections of the individual bipolar cells at cone terminals allow the transmission of information to the bipolar cells at two or three different levels. The synaptic ribbon complexes and synaptic connections of bipolar cells to photoreceptors outside synaptic ribbon complexes represent a first level of connections at the photoreceptors. The connections in the subsynaptic neuropil constitute a second level of neural connections at the photoreceptors. At least certain bipolar cells make synaptic connections in the common neuropil, furnishing a third level of neural connections.

The input of information to the information channels is contributed by several different types of neurons. The photoreceptors to which the bipolar cells are connected represent one type of neuron. In the case of bipolar cells connected to rods information is contributed both by rods and by cones. Large horizontal cells constitute a second type of neuron that transmits information regarding changes in the input received by horizontal cells from photoreceptors distributed over a large area of the retina. In contrast, small horizontal cells contribute information regarding the input from a relatively small number of photoreceptors confined to a small area of the retina located in one particular direction away from the information processing center, and they represent a third type of neuron contributing to the input to the information channel. Intrinsic bipolar cells constitute a fourth type of neuron among which we distinguished neurons that must contribute very different kinds of information. Thus, afferent and efferent bipolar cells are involved, and among the afferent bipolar cells the pattern of neural connections varies extensively, surely reflecting that different types of information is contributed to the different channels by these bipolar cells.

We can conclude that the input to the information processing center at the cone terminals involves a considerable amount of information. Whereas the horizontal cells transmit information horizontally within the outer circuitry layer, the bipolar cells transmit information both horizontally and vertically. Efferent bipolar cells, for instance, transmit information vertically.

With a bipolar cell connected to both cones and to rods, certain types of information processing can proceed at a low illuminance. For instance, bipolar cell 1, proposed to be involved in directionally selective responses, was connected both to cones and to rods. When the illuminance is too low to stimulate cones, the direct input from rods to the bipolar cell may allow information processing to proceed.

2. The Classification of Photoreceptors

Our discussion above did not link cone terminals to any particular spectral sensitivity of the photoreceptor. Instead, the shape of the terminal was linked to entirely different aspects of the photoreceptor function. Obviously, there is no reason why a small difference in the absorption spectrum of the photopigment should require the extensive difference in the structure between the rod and cone terminals.

The electron microscopic analysis revealing the structure of the cone terminals made it appear absurd to relate the cone-type terminals to the color discriminating function of cones. It was this absurdity that led Sjöstrand (1958) to propose that photoreceptors be classified on the basis of their synaptic terminals instead of on the basis of the shapes of the outer segments. The electron microscopic study at that time had already made it obvious that the structure of the retina could be analyzed at an entirely new level of refinement and that it was important that this analysis be pursued without being influenced by the dogmas developed from light microscopic study of the retina.

It was on this basis that Sjöstrand proposed that photoreceptors with the rod-type terminals be classified as α-photoreceptors and photoreceptors with cone-type terminals as β-photoreceptors. The guinea pig retina was at that time considered to be a pure rod retina both structurally and physiologically. Electrophysiologically, Granit (1947) had not been able to record any Purkinje shift from the guinea pig retina. Both α- and β-photoreceptors in the guinea pig retina should therefore be equipped with rhodopsin. In addition, β-photoreceptors would include conventional cones and this classification constituted a wider base for classification of cones than the generally accepted classification.

That rods can be mixed with photoreceptors with cone-shaped terminals in retinas lacking a Purkinje shift has in fact been shown electrophysiologically by recordings from a number of species of nocturnal animals (Dieterich, 1968). The objection can be raised that the electrophysiological recordings were not made with sufficient sensitivity to reveal a Purkinje shift, but this would mean that there would be few cones in these retinas, and it would be expected that the retinas would contain only a few photoreceptors with cone terminals. This is, however, not the case. There are numerous such photoreceptors in these retinas.

In the rat retina, Dodt and Echte (1961) reported two plateaus in the flicker fusion frequency, as revealed in the electroretinograms, and they interpreted this as evidence for the presence of two visual mechanisms. This observation was confirmed by Dowling (1967), who also showed that these two plateaus were not associated with any significant shift in the spectral sensitivity function. Rhodopsin therefore mediates both the fast and the slow flickering responses in the rat. This observation can be explained on the basis of our deduction in Chapter 7, Section 2.

Dowling also found that there are cone terminals in the rat retina, and he therefore made the interesting suggestion that "there may be cone-like receptors in the rat retina that have rhodopsin as their visual pigment." Obviously, these photoreceptors are Sjöstrand's β-photoreceptors!

Thus, there is something unsatisfying in the classification of photoreceptors when a structural feature that is considered to characterize cones is found in a photoreceptor equipped with rhodopsin, which according to Granit (1947, 1962) furnishes the most reasonable basis for classification of a rod.

Other structural features that have been considered to characterize cones, such as the conical shape of the outer segment and the presence of an oil droplet in the inner segment, apply only to certain types of cones. They are therefore not useful criteria for classification.

The extended analysis of the neural connections in the outer circuitry layer has supplied additional support for the concept that the shape of the terminals of the photoreceptors reflects basic aspects of the functional organization of the retina. A classification of the photoreceptors based on the morphology of the terminals is therefore even further justified. Fortunately, we can conserve the term "cone" because it describes perfectly the shape of the terminal of the β-photoreceptor! The cones must then be divided into chromatic and achromatic cones to account for differences in the spectral sensitivity of photoreceptors with conically shaped terminals. Achromatic cones are thus cones equipped with rhodopsin.

We end up distinguishing among rods, achromatic cones, and chromatic cones. This terminology has the obvious advantage over the earlier classification of being based on very basic physiological differences between rods and cones that apply to all cones. The differences between the roles the rods and cones play in establishing spatial brightness contrast enhancement and in information processing are much more basic than the shape of the outer segment and the presence of an oil droplet in the inner segment. The distinction between achromatic and chromatic cones precludes the routine conclusion that a photoreceptor that structurally qualifies as a cone is equipped with cone pigment and that the retina containing such photoreceptors therefore allows the discrimination of colors.

3. The Logic of Design

We have seen that structural information can be translated directly into a functional equivalent when the structural analysis has reached a level where it yields basic information. The discovery of the neuron as a structural unit in the nervous system represents a discovery at one such level.

The analysis of the neural connections at the rod and cone terminals reviewed here offers another example of structural features translated into their functional equivalents. As a result, the different types of responses of the bipolar cells and consequently of the ganglion cells could be deduced, and the presence of on-, off-, and on/off-responses could be explained. Even the detailed shapes of the curves revealing potential changes in the membrane of the bipolar cells as observed in electrophysiological recordings could be explained by the neural interactions at the photoreceptor terminals. The agreement between the deduced temporal changes in these responses and the observed changes shows that a translation of structure into function is possible.

The synaptic ribbon complex represents a fairly simple component of the circuitry of the retina involved in information processing. There are three information channels that converge onto the dendritic endings of one or two bipolar cells. The response of the bipolar cells, however, is not based on a simple summation of the input from these three channels.

Instead, the horizontal cells control or modulate the transmission between the photoreceptors and the bipolar cell. The gating function of the horizontal cells is regulated by the input to the horizontal cells at the synaptic ribbon complex as well as by the input within the entire functional receptive fields of the horizontal cells.

Neural interactions such as those deduced for the synaptic ribbon complex reveal features that have an enormous consequence for the analysis of information processing in the nervous system. It means that the responses of neurons are not based on a simple summation of the input from a certain number of neurons. As a consequence, information processing cannot be analyzed by recording responses from neurons. These recordings show only the final result of the integration of the input but not the way the input has been integrated at the level of that neuron. This is particularly critical because it is the latter integration that represents the primary event in information processing and, consequently, in the function of the brain.

An analysis of those primary events requires a detailed knowledge of the synaptic connections of the neurons, which can be achieved only through a three-dimensional reconstruction of the connections in neural centers. The analysis therefore depends on a direct translation of the function by a proper reading of the structure. The logic of the design of the nervous system makes this possible.

4. Some Generalizations

Certain aspects of the work presented here are likely to be of general significance, for instance, the connections of neurons according to particular patterns for certain types of information processing. The conditions for the deduction of the way information is processed at the level of primary events on the basis of the patterns of neural connections have therefore been satisfied.

Another concept that may be of general significance is the regulation of the membrane potential of a neuron by a depolarizing pressure opposed by a hyperpolarizing pressure. This mode of regulation seems to be required in the cases where a neuron responds with graded changes of the membrane potential. The lack of the threshold for stimulation, that characterizes neurons responding with spike potentials, eliminates conditions for these neurons to maintain a more or less fixed resting potential. Between discrete changes in the membrane potential these neurons will be characterized by a floating baseline potential.

The deduced gating action of neurons releasing a hyperpolarizing transmitter such as the horizontal cells is likely to be a basic feature of the circuitry of neural centers.

Neural circuitry may contain what we can refer to as standard parts, such as the standard parts in electronic circuits. The synaptic ribbon complex may well be one such standard part, although the location of the complex in an invagination of a neural membrane may be unique to the photoreceptors.

The dendritic tree is obviously the site for information processing with the dendritic endings picking up different pieces of information. For instance, in addition to connections to mixed synaptic ribbon complexes, bipolar cell 1 was connected to synaptic ribbon complexes at which both horizontal cells were large horizontal cells, directly to the photoreceptor outside synaptic ribbon complexes, to bipolar cells in the subsynaptic neuropil, and to small horizontal cells in the common neuropil away from the connections to the photoreceptor.

This means the picking up of both presynaptically processed information and primary information. The response of the bipolar cell will then be based on a weighing of the different types of transmitted information with possibilities for the end result to vary extensively. For instance, the responses of bipolar cell 1 will vary considerably depending upon the input from the small horizontal cells at the collaterals that we deduced can block signals picked up at the connections at the photoreceptor terminals.

Analysis of neural circuits allows conclusions regarding the basic events associated with information processing to be drawn. Such analysis requires that the neural connections be known in detail, not only as types of connections made by neurons identified incompletely by their shapes. Instead, the analysis requires that the connections of a certain population of neurons be revealed and that the neurons be classified on the basis of their patterns of connections to other neurons. This can only be done by a three-dimensional reconstruction of the neural connections. However, this structural analysis must be combined with an electrophysiological analysis to yield information regarding the electrical behavior of at least certain neurons in the circuit. These neurons can serve as check points in the circuits, guiding the deduction of how the circuits function.

For such an analysis of the nervous system it is necessary to invest much time. Structural analysis in particular will always be very time consuming. Spending this time, however, is justified because at present three-dimensional reconstruction is the only method that can contribute the information that is crucial for the analysis of the basic function of the brain, information processing.

The technique for reconstruction of neural connections in three dimensions has been improved greatly since work on which this book is based was done. A considerably larger volume of neural tissue can now be reconstructed and it would not be technically difficult to analyze, for instance, the circuitry across the entire retina, including both the outer and the inner plexiform layers.

The three-dimensional reconstruction of a large volume of neural tissue requires a very large number of electron micrographs. The time needed for the production of sufficient electron micrographs alone could be a whole year if a standard electron microscope is used. One type of electron microscope, however, was modified so that electron micrographs can be produced in only one-tenth to one-twentieth the time required with a standard electron microscope. On the basis of past experience, the tracing of the neural processes through the series of sections and the building of a model to store the information can now be pursued much more quickly.

With these technical improvements the challenging project of extending the analysis of the retinal circuitry to the entire circuits of both the outer and inner plexiform layers seemed a worthwhile goal. Unfortunately, when the technique, the equipment, and the experience made this possible, the work had to be stopped as a consequence of lack of support from granting agencies.

REFERENCES

Aguilar, M., and Stiles, W., *Opt. Acta* **1,** 59, 1954.

Allen, R. A., *in* The Retina, Morphology, Function, and Clinical Characteristics (Straatsma, B. R., Hall, M. O., Allen, R. A., and Crescitelli, F., Eds.), p. 10. Univ. of California Press, Berkeley/Los Angeles, 1969.

Arden, G. B., and Weale, R. A., *J. Physiol.* **125,** 417, 1954.

Baker, H. D., *J. Opt. Soc. Amer.* **43,** 798, 1953.

Baker, H. D., *J. Opt. Soc. Amer.* **53,** 98, 1963.

Baker, H. D., Doran, M. D., and Miller, K. E., *J. Opt. Soc. Amer.* **49,** 1065, 1959.

Barlow, H. B., and Hill, R. M., *Science* **139,** 412, 1963.

Barlow, H. B., and Levick, W. R., *J. Physiol. (London)* **170,** 53, 1964.

Barlow, H. B., and Levick, W. R., *J. Physiol. (London)* **178,** 477, 1965.

Barlow, H. B., Fitzhugh, R., and Kuffler, S. W., *J. Physiol. (London)* **137,** 338, 1957.

Barlow, H. B., Hill, R. M., and Levick, W. R., *J. Physiol. (London)* **173,** 377, 1964.

Baumgardt, E., *C. R. Seances Soc. Biol.* **143,** 786, 1949.

Baylor, D. A., *Fed. Proc.* **33,** 1074, 1974.

Baylor, D. A., Fuortes, M. G. F., and O'Bryan, P. M., *J. Physiol. (London)* **214,** 265, 1971.

Berger, C., *Acta Physiol. Scand.* **28,** 244, 1953.

Berger, C., *Acta Physiol. Scand.* **30,** 161, 1954.

Blackstad, T. W., *Z. Zellforsch. Mikrosk. Anat.* **67,** 819, 1965.

Bortoff, A. L., *Vision Res.* **4,** 626, 1964.

Boycott, B. B., and Dowling, J. E., *Phil. Trans. Roy. Soc. London B* **255,** 109, 1969.

Brown, J. E., *J. Neurophysiol.* **28,** 1091, 1965.

Byzov, A. L., *Cold Spring Harbor Symp. Quant. Biol.* **30,** 547, 1966.

Cervetto, L., and MacNichol, E. F., *Science* **178,** 767, 1972.

Cohen, A. I., *Handb. Sens. Physiol.* **VII/2,** 63, 1972.

Crawford, B. H., *Proc. Roy. Soc. London B* **134,** 283, 1947.

Daw, N. W., *J. Physiol. (London)* **197,** 567, 1968.

de Robertis, E., and Bennett, H. S., *Fed. Proc.* **13,** 35, 1954.

de Robertis, E., and Bennett, H. S., *J. Biophys. Biochem. Cytol.* **1,** 47, 1955.

Detwiler, S. R., *Exp. Biol. Monogr., Vertebrate Photoreceptors, New York,* 1943.

Dieterich, C. E., *Albrecht von Graefes Arch. Klin. Exp. Ophthalmol.* **174,** 289, 1968.

Ditchburn, R. W., and Ginsburg, B. L., *Nature (London)* **170,** 36, 1952.

Dodt, E., and Echte, K., *J. Neurophysiol.* **24,** 427, 1961.

Dowling, J. E., *J. Gen. Physiol.* **46,** 1287, 1963.

Dowling, J. E., *Science* **157,** 584, 1967.

Dowling, J. E., *Proc. Roy. Soc. London B* **170,** 205, 1968.

Dowling, J. E., and Werblin, F. S., *Vision Res. Suppl.* **3,** 1, 1971.

Fain, G. L., Gold, G. H., and Dowling, J. E., *Cold Spring Harbor Symp. Quant. Biol.* **40,** 547, 1976.

Gallego, A., *Vision Res. Suppl.* **3,** 33, 1971.

Gay, H., and Anderson, T. D., *Science* **120,** 1071, 1954.

Gerschenfeld, H., and Piccolini, M., *Nature (London)* **268,** 257, 1977.

Gouras, P., *J. Physiol. (London)* **152,** 487, 1960.

Graham, C. H., Vision and Visual Perception. John Wiley and Sons, New York, 1966.

Granit, R., Sensory Mechanisms of the Retina. Geoffrey Cumberlege, Oxford Univ. Press, 1947.
Granit, R., *in* The Eye (Davson, H., Ed.), Vol. 2, p. 575. Academic Press, New York, 1962.
Greff, R., *Graefe-Saemischs Handb. Ophthal.* **1/1,** Chap. 5, 1900.

Haig, C., *in* Visual Problems of Colour, National Physical Laboratory Symposium, No. 8, Vol. I, p. 192. H. M. Stationary Office, London, 1958.
Hartline, H. K., *Amer. J. Physiol.* **121,** 400, 1938.
Hartline, H. K., *Amer. J. Physiol.* **130,** 690, 1940.
Hartline, H. K., *Fed. Proc.* **8,** 69, 1949.
Hartline, H. K., and Ratliff, F., *J. Gen. Physiol.* **40,** 357, 1957.
Hecht, S., and Shlaer, S., *J. Gen. Physiol.* **19,** 956, 1936.
Hecht, S., and Shlaer, S., *J. Opt. Soc. Amer.* **28,** 269, 1938.
Hecht, S., and Smith, E. L., *J. Gen. Physiol.* **19,** 979, 1936.
Hecht, S., and Verrijp, C. D., *J. Gen. Physiol.* **17,** 251, 1933.
Hecht, S., Haig, C., and Chase, A. M., *J. Gen. Physiol.* **20,** 831, 1937.
Hubel, D. H., *J. Physiol. (London)* **147,** 226, 1959.
Hubel, D. H., and Wiesel, T. N., *J. Physiol (London)* **148,** 574, 1959.
Hubel, D. H., and Wiesel, T. N., *J. Physiol. (London)* **195,** 215, 1968.

Kaneko, A., *J. Physiol (London)* **207,** 623, 1970.
Kaneko, A., *J. Physiol. (London)* **213,** 95, 1971.
Kaneko, A., *J. Physiol. (London)* **235,** 133, 1973.
Kaneko, A., and Hashimoto, H., *Vision Res.* **9,** 37, 1969.
Karwoski, G. I., and Burkhardt, D. A., *Vison Res.* **16,** 1483, 1976.
Kidd, M., *in* Cytology of Nervous Tissue, Anat. Soc. Great Britain and Ireland, p. 88. Taylor & Francis, London, 1961.
Kidd, M., *J. Anat.* **96,** 179, 1962.
Kolb, H., *Phil. Trans. Roy. Soc. London B* **258,** 261, 1970.
Kuffler, S. W., *Cold Spring Harbor Symp. Quant. Biol.* **17,** 281, 1952.
Kuffler, S. W., *J. Neurophysiol.* **16,** 37, 1953.

Leicester, J., and Stone, J., *Vision Res.* **7,** 695, 1967.
Lettvin, J. Y., Maturana, H. R., McCulloch, W. S., and Pitts, W. H., *Proc. Inst. Radio Eng.* **47,** 1940, 1959.
Lettvin, J. Y., Pitts, W. H., and McCulloch, W. S., *in* Sensory Communication. MIT Press/Wiley, New York/London, 1961.
Levick, W. R., *J. Physiol. (London)* **188,** 285, 1967.
Levick, W. R., *in* Handbook of Sensory Physiology, Vol. VII/2, p. 531, 1972.

Mach, E., *Sitzungsber. Math. Naturwiss. Classe, Kaiserl. Akad. Wiss. Wien* **52,** 302, 1865.
Maksimova, E. M., *Biofizika* **14,** 537, 1969.
Matsumoto, N., and Naka, K.-I., *Brain Res.* **42,** 59, 1972.
Maturana, H. R., and Frenk, G., *Science* **142,** 977, 1963.
Maturana, H. R., Lettvin, J. Y., McCulloch, W. S., and Pitts, W. H., *J. Gen. Physiol.* **43,** Suppl. 11, 129, 1960.
Michael, C. R., *J. Neurophysiol.* **31,** 249, 1968.
Missotten, L., The Ultrastructure of the Human Retina. Arscia/Uitgaven, Brussels, 1965.
Missotten, L., Appelmans, M., and Michiels, J., *Bull. Mem. Soc. Fr. Ophthalmol.* **76,** 59, 1963.

Naka, K.-I., *Science* **171,** 691, 1971.
Naka, K.-I., and Nye, P. W., *J. Neurophysiol.* **33,** 625, 1970.
Naka, K.-I., and Ohtsuka, T., *J. Neurophysiol.* **38,** 72, 1975.
Naka, K.-I., and Rushton, W.A.H., *J. Physiol. (London)* **192,** 437, 1967.
Normann, R. A., and Pochobradsky, J., *J. Physiol. (London)* **261,** 15, 1976.
Norton, A. L., Spekreijse, H., Wagner, H. G., and Wolbarsht, M. L., *J. Physiol. (London)* **206,** 93, 1970.

O'Daly, J. A., *Nature (London)* **216,** 1329, 1967.
Oyster, C. W., *J. Physiol. (London)* **199,** 613, 1968.

Pirenne, M. H., *in* The Eye (Davson, H., Ed.), Vol. 2, Chapter 5, p. 93. Academic Press, New York, 1962.
Polyak, S. L., The Retina. Chicago Univ. Press, Chicago, 1941.
Pütter, A., *Graefe-Saemischs Handb. Ges. Augenheilk.* **2/1,** Chap. X, 1, 1912.

Ramon y Cajal, S., *Cellule* **9,** 121, 1892 (translated to English in R. W. Rodieck, The Vertebrate Retina, Freeman, San Francisco, 1933).
Ramon y Cajal, S., Die Retina der Wirbeltiere, Wiesbaden, 1894.
Ramon y Cajal, S., *Trav. Lab. Rech. Biol.,* Univ. Madrid, 1933.
Ratliff, F., Mach Bands: Quantitative Studies on Neural Networks in the Retina. Holden–Day, San Francisco, 1965.
Raviola, E. W., and Gilula, N. B., *Proc. Natl. Acad. Sci. USA* **70,** 1677, 1973.
Riggs, L. A., Ratliff, F., Cornsweet, J. C., and Cornsweet, T. N., *J. Opt. Soc. Amer.* **43,** 495, 1953.
Riggs, L. A., Ratliff, F., and Keesey, U. T., *J. Opt. Soc. Amer.* **51,** 702, 1961.
Rochon-Duvigneaud, A., Recherches sur l'Oeil et la Vision chez les Vertébrés. Laval, 1934.
Rodieck, R. W., The Vertebrate Retina. Freeman, San Francisco, 1973.
Rodieck, R. W., and Rushton, W. A. H., *J. Physiol.* **254,** 775, 1976.
Rushton, W. A. H., and Powell, D. S., *Vision Res.* **12,** 1083, 1972.

Schultze, M., *Arch. Mikrosk. Anat.* **2,** 175, 1866.
Sjöstrand, F. S., *J. Appl. Phys.* **24,** 117, 1953a.
Sjöstrand, F. S., *J. Appl. Phys.* **24,** 1522, 1953b.
Sjöstrand, F. S., *Experientia* **9,** 68, 1953c.
Sjöstrand, F. S., *Z. Wiss. Mikrosk.* **62,** 65, 1954a.
Sjöstrand, F. S., *Z. Wiss. Mikrosk.* **62,** 79, 1954b.
Sjöstrand, F. S., *in* Fine Structure of Cells, Symp. Congr. Cell Biol. 8th Leyden, 1954, p. 222. Nordhoff, Groningen, 1955.
Sjöstrand, F. S., *Proc. Int. Conf. Electron Microsc. 3rd* **1954,** 428, 1956.
Sjöstrand, F. S., *J. Ultrastruct. Res.* **2,** 122, 1958.
Sjöstrand, F. S., *Ergebn. der Biologie* **21,** 128, 1959.
Sjöstrand, F. S., CIBA Symp. on Physiology and Experimental Psychology of Colour Vision (Wolstenholme, G. E. W., and Knight, J., Eds.). Churchill, London, 1965.
Sjöstrand, F. S., *in* The Retina: Morphology, Function and Clinical Characteristics (Straatsma, B. R., Hall, M. O., Allen, R. A., and Crescitelli, F., Eds.), p. 63. Univ. of California Press, Los Angeles, 1969.
Sjöstrand, F. S., *J. Ultrastruct. Res.* **49,** 60, 1974.
Sjöstrand, F. S., *Vision Res.* **16,** 1, 1976.
Sjöstrand, F. S., *J. Ultrastruct. Res.* **62,** 54, 1978.

Sjöstrand, F. S., and Nilsson, S. E., *in* The Rabbit in Eye Research (Prince, J. H., Ed.) p. 449. Charles C. Thomas, Springfield, 1969.
Smith, C. A., and Sjöstrand, F. S., *J. Ultrastruct. Res.* **5,** 184, 1961.
Stell, W. K., *Anat. Rec.* **153,** 389, 1965.
Stell, W. K., *Amer. J. Anat.* **121,** 401, 1967.
Svaetichin, G., *Acta Physiol. Scand.* **29,** Suppl. 106, 565, 1953.

Trifinov, Yu. A., *Biofizika* **13,** 809, 1968.
Trifinov, Yu. A., and Ostrovskii, M. A., *Neirofiziologiya* **2,** 79, 1970.
Tuttle, J. R., *Vision Res.* **17,** 777, 1977.

Wald, G., *Science* **119,** 887, 1954.
Wald, G., *in* Symposium: Visual Problems of Color, p. 15. Chemical Publishing Co. New York, 1961.
Weale, R. A., *Nature (London)* **181,** 154, 1958.
Weale, R. A., *Ann. Roy. Coll. Surg.* **28,** 16, 1961.
Werblin, F. S., *J. Neurophysiol.* **33,** 342, 1970.
Werblin, F. S., and Dowling, J. E., *J. Neurophysiol.* **32,** 339, 1969.
Wyatt, H. J., and Daw, N. W., *J. Neurophysiol.* **38,** 613, 1975.

Yamada, E., and Ishikawa, T., *Cold Spring Harbor Symp. Quant. Biol.* **30,** 383, 1965.

Author Index

H

I

K

L

M

N

O

P

R

S

T

Subject Index

N

O

P

R